SYMBOLIC ANALYSIS FOR AUTOMATED DESIGN OF ANALOG INTEGRATED CIRCUITS

THE KLUWER INTERNATIONAL SERIES IN ENGINEERING AND COMPUTER SCIENCE

VLSI, COMPUTER ARCHITECTURE AND DIGITAL SIGNAL PROCESSING

Consulting Editor

Jonathan Allen

Latest Titles

SYMBOLIC ANALYSIS FOR AUTOMATED DESIGN OF ANALOG INTEGRATED CIRCUITS

by

Georges Gielen
Katholieke Universiteit Leuven

and

Willy Sansen
Katholieke Universiteit Leuven

Kluwer Academic Publishers
Boston/Dordrecht/London

Distributors for North America:
Kluwer Academic Publishers
101 Philip Drive
Assinippi Park
Norwell, Massachusetts 02061 USA

Distributors for all other countries:
Kluwer Academic Publishers Group
Distribution Centre
Post Office Box 322
3300 AH Dordrecht, THE NETHERLANDS

Library of Congress Cataloging-in-Publication Data

Gielen, Georges.
    Symbolic analysis for automated design of analog integrated circuits / by
Georges Gielen and Willy Sansen.
        p.   cm. — (The Kluwer international series in engineering and computer
science ; SECS 137)
    Includes bibliographical references and index.
    ISBN 0-7923-9161-6
    1. Linear integrated circuits—Design and construction—Data processing.
2. Electronic circuit design—Data processing. 3. Computer—aided design. I.
Sansen, Willy M.C. II. Title. III. Series.
TK7874.G54   1991
621.381'5—dc20   91-15032
                                                                          CIP

Copyright © 1991 by Kluwer Academic Publishers

All rights reserved. No part of this publication may be reproduced, stored in a retrieval
system or transmitted in any form or by any means, mechanical, photo-copying, recording,
or otherwise, without the prior written permission of the publisher, Kluwer Academic
Publishers, 101 Philip Drive, Assinippi Park, Norwell, Massachusetts 02061.

Printed on acid-free paper.

Printed in the United States of America

TABLE OF CONTENTS

LIST OF SYMBOLS AND ABBREVIATIONS

A/D	: analog/digital
ADC	: analog-to-digital converter
A_o	: differential-mode gain
ASIC	: application-specific integrated circuit
CAD	: computer-aided design
C_C	: compensation capacitor
C_i	: node capacitance
C_L	: load capacitor
CMNA	: compacted modified nodal analysis
CMRR	: common-mode rejection ratio
D/A	: digital/analog
DA	: design automation
DAC	: digital-to-analog converter
DSLE	: double sparse Laplace expansion
DSLEM	: double sparse Laplace expansion with memo storage
DSP	: digital signal processor
ε_A	: accumulated absolute error
ε_N	: accumulated effective error (or nominal error)
f	: frequency
FDSLEM	: factoring double sparse Laplace expansion with memo storage
GBW	: gain-bandwidth
HD2	: second harmonic distortion ratio
HD3	: third harmonic distortion ratio
MNA	: modified nodal analysis
NA	: nodal analysis
opamp	: operational amplifier
OTA	: operational transconductance amplifier
$\underline{p}$	: vector of symbolic circuit parameters
p_i	: ith pole

PM   : phase margin
PSRR   : power-supply rejection ratio
R_L : load resistor
s   : complex frequency variable of the Laplace transform
SC   : switched capacitor
SLE   : sparse Laplace expansion
SLEM   : sparse Laplace expansion with memo storage
SLEMP   : sparse Laplace expansion with memo storage and preordering
SLEP   : sparse Laplace expansion with preordering
SOP   : sum of products
SR   : slew rate
T_s : settling time
VLSI   : very large scale integration
VLSIC   : very large scale integrated circuit
V_{os} : offset voltage
ω   : pulsation
x   : complex frequency variable
z   : complex frequency variable of the z-transform
z_i : ith zero

Notational conventions

variables are denoted in lowercase

vectors are denoted in underlined lowercase

MATRICES are denoted in uppercase

All examples are ended by ◆

FOREWORD

It is a great honor to provide a few words of introduction for Dr. Georges Gielen's and Prof. Willy Sansen's book *"Symbolic analysis for automated design of analog integrated circuits"*. The symbolic analysis method presented in this book represents a significant step forward in the area of analog circuit design. As demonstrated in this book, symbolic analysis opens up new possibilities for the development of computer-aided design (CAD) tools that can analyze an analog circuit topology and automatically size the components for a given set of specifications. Symbolic analysis even has the potential to improve the training of young analog circuit designers and to guide more experienced designers through second-order phenomena such as distortion. This book can also serve as an excellent reference for researchers in the analog circuit design area and creators of CAD tools, as it provides a comprehensive overview and comparison of various approaches for analog circuit design automation and an extensive bibliography.

The world is essentially analog in nature, hence most electronic systems involve both analog and digital circuitry. As the number of transistors that can be integrated on a single integrated circuit (IC) substrate steadily increases over time, an ever increasing number of systems will be implemented with one, or a few, very complex ICs because of their lower production costs. As a result of these trends, there has been and will continue to be a steady increase in the number of mixed analog-digital ICs designed. One recent study of The Technology Research Group, Inc., reported that in 1990 approximately 60% of all CMOS and BiCMOS Application Specific IC (ASIC) designs currently incorporate analog circuit modules. Designing the analog portion of a mixed analog-digital IC is frequently the bottleneck that limits one of the most important economic aspects of ASIC manufacturing: time-to-market. Therefore, to decrease the required design time for the analog circuitry, more electrical engineers skilled in analog circuit design must be trained, and more sophisticated computer-aided analog design tools must be developed.

Symbolic analysis has the potential to offer insight to students and practitioners of analog circuit design. Too many circuit designers rely only on multiple numerical circuit simulations (e.g. SPICE) to provide insight into the behavior of a circuit - symbolic analysis can provide a much richer understanding of the behavior of a given analog circuit topology.

Recent successes in automating the design of basic analog building blocks such as opamps and comparators have almost uniformly employed an "equation-based" approach that substitutes analysis equations for simulation in order to predict the performance of an analog circuit. Unfortunately, these analysis equations are circuit topology specific and their development normally requires a great deal of time and effort on the part of expert analog circuit designers. The analog design problem is compounded by the fact that to respond efficiently to a broad range of applications usually requires that a wide variety of analog circuit topologies be available. As this book demonstrates, symbolic analysis can be used to automatically generate a significant fraction of the analysis equations needed to characterize a new circuit topology. Therefore, symbolic analysis is an important step forward in the development of CAD tools that aid in analog circuit design.

L. Richard Carley

Department of Electrical and Computer Engineering
Carnegie Mellon University
Pittsburgh, PA USA

PREFACE

Analog design automation is a field of increasing interest in industry. Analog circuits are demanded for interfacing as well as for many high-performance applications. At the same time, economical reasons drive ASICs into integration of complete mixed analog-digital systems on one chip. In both applications, however, the lack of analog CAD tools results in long design times and costly design errors. This forms a major obstacle for the realization of completely integrated mixed-signal systems, fully utilizing the potential of the technology. This explains the large industrial demand for analog design automation tools nowadays. Accordingly, many research activities are going on in this emerging field. Our group, the ESAT-MICAS division of the Katholieke Universiteit Leuven, Belgium, was one of the first to start with this research [DEG_84]. Then, since about 1985, we have noticed a strong increase worldwide in the analog CAD research, resulting now in the first prototype programs for the automated sizing and layout of (mostly lower-level) analog circuits.

Recently, we also have initiated some courses at our university about this analog design automation. However, when looking through the literature, we cannot find any general text which adequately describes the major aspects and the state of the art in the field, and which presents a broad overview and general comparison of the different approaches. This book was written to fill this gap. Historically, it has evolved from the PhD dissertation of the first author [GIE_90d], which describes the research carried out in the ESAT-MICAS group. As such, the book has two purposes. Firstly, it provides a general introduction to analog design automation, points out the major areas and discusses the major realizations and difficulties. Secondly, it focuses in more detail on one particular aspect: the symbolic simulation of analog integrated circuits, in which we have built up a strong experience. Symbolic simulation is indeed of increasing interest to analog designers for it provides both experienced and inexperienced designers with insight into the behavior of a circuit. This is especially true for more complicated characteristics such as the PSRR and the harmonic distortion at higher frequencies and in the presence of mismatches. In addition, as will be described in this book, symbolic analysis techniques can be used to create an open analog design system in which the inclusion of new circuit topologies is highly simplified.

The book addresses both practicing designers and CAD developers. It therefore tries to keep a good balance between general introduction, applications and examples on one hand, and still providing enough algorithmic details on the other hand. We believe that we have succeeded in this goal and that both groups of readers will find enough useful material in this book. At the same time, the book can also be used for advanced graduate courses in electrical engineering, more particular in CAD courses covering analog CAD.

Finally, we have to express our gratitude to all persons who contributed towards the realization of this book. In particular, we wish to thank prof. Hugo De Man and prof. Joos Vandewalle, who proofread the thesis and made many useful comments. We also want to thank Koen Swings, Herman Walscharts and Piet Wambacq, who cooperated on analog CAD research in our group. Finally, we are also grateful to the Belgian National Fund of Scientific Research and to the Philips Research Laboratories, Eindhoven, The Netherlands, for logistic support of our research.

This book is dedicated to Patricia and Hadewych.

Georges Gielen
Willy Sansen

Department of Electrical Engineering
Katholieke Universiteit Leuven
Leuven, Belgium

SYMBOLIC ANALYSIS FOR AUTOMATED DESIGN OF ANALOG INTEGRATED CIRCUITS

1

INTRODUCTION TO
ANALOG DESIGN AUTOMATION

1.1. Introduction

Advances in VLSI technology nowadays allow the realization of complex integrated electronic circuits and systems. Application-specific integrated circuits (ASICs) are moving towards the integration of complete systems. These microsystems include both digital and analog parts on a single chip. At the same time, the use of computer-aided design (CAD) tools has become indispensible to reduce the total design time and cost. For digital VLSI, several methodologies and design tools have been developed in the past, resulting in complete digital silicon compilers [GAJ_88]. The first analog design tools and methodologies, however, are just now being introduced [DEG_87c, GIE_90a, HAR_89b, KOH_90, etc.]. Most of the design is still carried out manually by analog experts. These experts derive simplified circuit expressions by hand, iteratively use numerical circuit simulators by trial and error and handcraft the layout. This explains why the analog part - though usually small in area - takes more and more of the overall design time and cost of the present mixed analog-digital chips. This is especially true in high-performance applications, requiring for example high-frequency performance, high precision, low noise and/or low distortion. Besides, due to the hand design and hand layout, errors frequently occur in the small analog section, leading to several reruns. This delay cannot be tolerated in most applications where the time to market is critical for profit.

The reasons for this present lack of analog design automation tools can be found in the nature of analog design itself. Designing high-performance analog circuits is a very knowledge-intensive and complicated task because of the many (functional and parasitic) interactions and trade-offs to be managed. The quality (and often the functionality) of the resulting design strongly relies on the insight and expertise of the analog designer. This expert knowledge though is quite intuitive, poorly structured and far from generally accessible in a knowledge base. On the other hand, digital design allows for more modularity, design abstraction and hierarchy. All this makes the automation of the analog design process to a very complicated task. It also explains why analog CAD nowadays is far behind its digital counterpart and, to some large extent, also why integrated electronics nowadays is characterized by a high digitalization.

Nevertheless, analog circuits are still required in many applications. Signals in the real world are analog in nature and all interface circuitry (such as analog-to-digital and digital-to-analog converters) then necessarily contain some analog parts. In addition, analog still outperforms digital for various high-performance applications, such as high-frequency circuits for cellular mobile telephony or low-noise data-acquisition systems for sensor interfaces or particle/radiation detectors.

This inevitable need for analog circuits and the economics-driven tendency in ASIC design to integrate total mixed analog-digital systems onto a single chip has created the present large industrial demand for analog computer-aided design (CAD) and design automation (DA) tools. This book therefore presents the state of the art in this increasingly important field. The basic concepts are illustrated by discussion of a flexible methodology for the automated design of analog functional modules (or building blocks), such as operational amplifiers, comparators, filters or converters. The key aspect of this methodology is that it uses symbolic simulation and design equation manipulation as a solution to build an open analog synthesis system, in which the inclusion of new circuit schematics and design knowledge is highly simplified. This then provides the motivation for the detailed discussion of symbolic simulation which is the major topic of this book. Both the functional and the algorithmic aspects of the symbolic simulation of analog circuits are described in great detail. The present state of the art in this field, which is of increasingly growing interest to the design community, is also depicted. Finally, a design example shows how to use the symbolic analysis techniques to put new knowledge into the synthesis system and to apply this knowledge to size practical circuits.

At this point, it is important to emphasize that this text focuses on the automated design of integrated circuits and not on board-level design. Up till now, however, no real analog design tools are available at the level of an integrated (most likely mixed analog-digital) system. **This book therefore explains about analog symbolic simulation and its use in a methodology for the automated design of analog integrated modules (or functional blocks).**

The context of the book is explained in this first chapter which presents a general introduction into analog design automation. Firstly, in the next section 1.2, some important definitions with respect to analog design are presented. These definitions are extensively used throughout the remainder of this text. Section 1.3 briefly reviews the characteristics of analog design and points out the similarities and differences with digital design in order to provide some feeling for the complexity of the analog design automation task. The application domains for analog integrated circuits are then described in section 1.4. This results in the formulation of the goals of analog design automation. Section 1.5 then summarizes the different approaches or styles used to design analog circuits from firmware over semi-custom to full-custom design techniques. For most high-performance applications, however, the analog circuits must be tailored and tuned to the actual application. This leads to an analog silicon compiler which automates the design process from system specifications down to layout. A methodology outline for the synthesis of analog or mixed analog-digital integrated systems is then presented in section 1.6. It divides the overall design task in system design and module design. Section 1.6 also shortly describes the tools needed at the system level. The methodology for the synthesis of the analog modules will then be presented in the next chapter. Finally, section 1.7 gives the general outline of this book.

1.2. Definitions in analog design automation

As analog design automation is a rather new research field, with many different approaches and try-outs currently being published, there still is large confusion and disagreement about the applied terminology. Therefore, in this section, we provide some important definitions, which are extensively used throughout the remainder of this book. These definitions have to clarify the terminology to the reader and are stated generally enough to serve as a reference base for future publications in this field.

1.2.1. Definitions of analog design and analog design automation

The overall context of this book is the automation of the design of analog integrated circuits, both time-continuous and time-discrete (e.g. switched-capacitor). These analog circuits are usually part of larger, more complicated VLSI systems, which are integrated on a single chip and which most often also contain digital circuits. This leads us to the following general definition of analog design.

Definition: analog design
Design is the translation of a behavioral system description into an interconnection of circuit schematics (structure) and physical layout (silicon) over several levels of hierarchy.

For analog (sub)systems, the behavioral description describes what the (sub)system has to do (not how to do it!) and what the (sub)system's performance specifications are. Design is then the translation of this description into structure and layout such that the resulting silicon after processing performs the specified behavior. To give an idea of the kind of systems and complexity we are talking about, consider the following example of a transmitter chip for a full-duplex modem.

Example
Fig. 1.1 shows the simplified principle schematic for the use of a modem. The modem forms the interface between a digital data source/receiver, such as a personal computer, and the telephone line, and consists of a transmitter and a receiver part. The transmitter chip has to transform the digital bits at its input into the appropriate

analog waveforms to be sent over the telephone line according to some CITT norm with a given baud rate, accuracy, dynamic range, etc. (e.g. 9600 bits/s, 0.1% accuracy, less than -60 dB in-band noise and distortion). It contains functions such as modulation, filtering, line equalization and line driving. The receiver chip then performs the inverse operation.   ◆

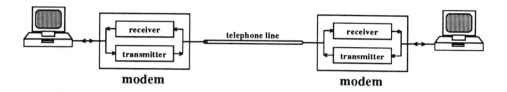

Fig. 1.1. Simplified principle schematic of the use of a modem to illustrate the complexity of present-day integrated-system designs.

     With respect to automating analog design, the two following definitions are important.

Definition: computer-aided design (CAD)

Computer-aided design tools are computer programs, such as a simulator or a symbolic layout editor, which support the designer in some particular design task in order to make design easier.

Definition: design automation (DA)

Design automation tools are tools which automate some design tasks, such as automatic sizing and circuit optimization, and which are normally integrated into one global software system covering the whole design path from behavioral system description to layout.

     The above distinction between CAD tools and DA tools, however, is somehow artificial, since many tools are used both interactively to support human designers and as part of an integrated design automation system. A typical example is a numerical simulator, such as SPICE [VLA_80b], which designers use to verify their circuit designs. The same simulator is used in an automated design system as well, but the automatic verification of say an operational amplifier requires many different simulations as well as other tests to be carried out.

     After these definitions of analog design and analog design automation, the following subsection investigates the need for introducing hierarchy during the automated design of analog or mixed analog-digital systems.

1.2.2. Importance of hierarchy

The automated design of analog or mixed analog-digital systems, such as the above modem chip, starting from their behavioral description is a very complex task. To make it feasible, the design task has to be split up into smaller subtasks, which are then again split up into smaller subsubtasks, and so forth. This is the well-known divide-and-conquer strategy. The design methodology thus has to be structured over several levels of hierarchy. Each hierarchical level then corresponds to a different level of design abstraction and has different design primitives to manipulate.

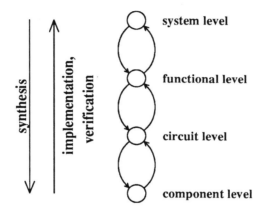

Fig. 1.2. Hierarchical organisation of analog design.

For analog CAD and DA, a plausible **hierarchical organisation**, as shown in Fig. 1.2, could be :
- *system (chip) level* : with the system as design primitive e.g. modem chip, data-acquisition chip...
- *functional level* : with analog functional modules as primitives e.g. filter, A/D converter, phase-locked loop...
- *circuit level* : with circuits as primitives e.g. operational amplifier, comparator, oscillator...
- *component level* : with components as primitives e.g. transistor, capacitor, resistor...

Note, however, that this hierarchical organisation for analog design is rather functional and that it is hard to formalize, as will be discussed in section 1.3. For the above example of the modem transmitter chip, the hierarchical decomposition is illustrated in Fig. 1.3. At top level, the transmitter contains a filter as one of the functional modules. This filter then consists of one or more operational amplifiers, which are themselves composed of transistors.

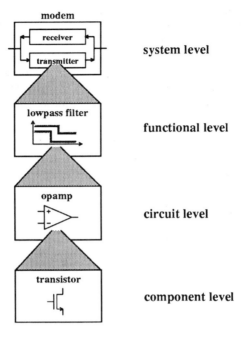

Fig. 1.3. Example of hierarchical decomposition for a modem transmitter chip.

Note also that for analog designs the hierarchical levels are not strictly defined. There can be several sublevels in between the above levels. For example, an *n*th-order filter can consist of a cascade of biquadratic sections, each of which consists of operational amplifiers and passive elements. An operational amplifier can be divided into subcircuits such as differential pairs, current mirrors, etc. which consist themselves of transistors and passive elements. Moreover, there is no general agreement yet in the analog design community about the actual hierarchical levels to be used for analog design automation.

The design of any system then proceeds by hierarchically decomposing the system into an interconnection of smaller and smaller blocks at lower abstraction levels whereby gradually more and more circuit detail is created, and by translating the system-level specifications down in this hierarchy until a level is reached which allows a physical implementation (this is the component level for full-custom design). This process is called top-down synthesis.

Definition: synthesis
Synthesis is the translation of a system under design from a higher level in the hierarchy down to a lower level.

Synthesis is one part of the total system design process. It is accompanied by a bottom-up physical implementation and verification process. The physical layouts of all lower-level cells are assembled together by placement and routing in order to obtain the physical implementation of the total system. Also, it has to be verified that no errors have been made during the hierarchical translation and that no local errors occur at each hierarchical level, in order to guarantee that the final physical implementation of the system satisfies the specified system-level behavior. This results in the following definition.

Definition: verification
Verification is proving that a lower-level representation of the system under design has the same behavior as the higher level(s), and that it is functionally correct at each level.

The verification can be accomplished with simulation, knowledge-based and/or formal techniques. Verification, however, is redundant if it can be guaranteed during top-down synthesis that any higher-level block satisfies its performance specifications if all its lower-level blocks satisfy their translated performance specifications. But proving this is even more difficult than simply verifying the design.

The total hierarchical design process can now be indicated on Fig. 1.2. The design of a system proceeds by synthesis from a higher level to a lower level down in the hierarchy until a level is reached which allows a physical implementation. The design is then completed by moving again up in the hierarchy, assembling all lower-level layouts and verifying all previous synthesis steps. Hence, if we want to automate the analog design process, automatic synthesis and verification tools have to be present at all hierarchical levels considered in the design strategy. It is however important to realize that Fig. 1.2 only conceptually illustrates the ideas of hierarchy and top-down synthesis and bottom-up implementation/verification. Several hierarchical design methodologies exist which implement the actual design process in practice. We will present one such a methodology in section 1.6.

After this introduction of hierarchical levels, we also have to describe the different views to represent a system and the corresponding terminology.

1.2.3. From behavior over structure to geometry

At each level of the hierarchy used in the automated design strategy, the system or circuit under design can be represented by **three different views** :
- a *behavior view* : behavioral or functional description of each block at the given hierarchical level (*"what to do"*).
- a *structure view* : description of how each block is composed of lower-level blocks (*"how to do it"*)
- a *geometry view* : physical implementation of each block (*layout*).

These views were originally recognized in the digital domain [GAJ_83], but of course hold for analog as well. For example, the three views of an operational amplifier are shown in Fig. 1.4. Note that the behavioral view is supplemented with constraint specifications (performance constraints, environmental constraints, test constraints).

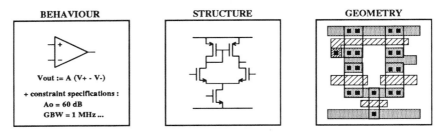

Fig. 1.4. Behavior, structure and geometry view of an operational amplifier.

Design is then the translation of behavior over structure into geometry. For analog modules, this design process can conceptually be decomposed into several subtasks, as depicted in Fig. 1.5. This is also the design sequence followed by most human designers. The behavioral description for the module consists of the behavior (function) to be realized by the module, and is supplemented with the corresponding performance and environmental specifications. This behavior is described in an analog behavioral language (for example AHDL [KUR_90]).

1. The translation of the behavioral description into structure is called **schematic synthesis**. This translation is performed in two steps.
   - First, a topology has to be generated, which is able to realize the function and the specifications. This step is called **topology generation** and yields an unsized schematic. A **topology** is defined as an interconnection of subblocks from a lower level in the hierarchy. For example, the topology for a generic two-stage operational amplifier is shown in Fig. 1.6. The subblocks are a

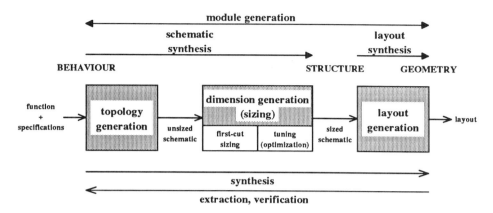

Fig. 1.5. Block diagram illustrating the subtasks and the terminology for the automatic generation of analog modules.

differential pair, a current mirror, two current sources, a gain stage, a level shifter and a compensation stage. Only at the lowest level, the subblocks are transistors and passive components. As the hierarchy thus corresponds to a functional decomposition of blocks into smaller subblocks, the number of hierarchical levels in topology generation depends on the type of the module. For an opamp for instance, three levels can be distinguished: the opamp level, the subcircuit level and the component level [HAR_87]. For higher-level modules, such as A/D and D/A converters, more hierarchical levels have to be distinguished.

- Secondly, the parameter values (e.g. device sizes) of the previously generated topology are determined. This step is called **sizing or dimensioning or dimension generation** and yields a sized schematic. This sizing is often performed in two phases: a fast first-cut design followed by some fine-tuning (often by means of optimization routines).

2. The translation of this structure into geometry is then called **layout generation or layout synthesis**. It yields the layout of the module.

The overall top-down translation process is called **module synthesis** and has to be completed by the bottom-up process of layout extraction and module verification. The total process of automatically designing an analog module by topology generation, sizing, layout generation and verification is called **analog module generation**. Note that this is similar to digital module generation, which, however, usually puts more emphasis on layout and less on sizing than analog module generation. Also, it is important to realize that Fig. 1.5 only presents a conceptual decomposition of analog module generation in order to define the terminology. Several methodologies exist that implement automated analog module generation in practice, but they all include the above subtasks in one way or another. These practical systems will be discussed in the next chapter.

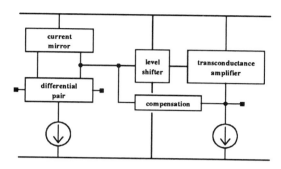

Fig. 1.6. Generic topology for a two-stage operational amplifier.

It is also important here to distinguish **two types of designs** :
- the first category are the designs at the edge of the performance limitations (as dictated by the technology). These designs require new circuit schematics, new tricks, careful design and layout in order to meet the stringent specification requirements. Hence, these designs will never be fully automated and will always need to be finished by expert designers.
- the second category are the designs with more relaxed, moderate specifications which can be designed quite easily with the existing schematics and the existing knowledge. Hence, these designs can be automated and will be performed by an automated analog design system in the future.

Analog design automation will certainly not displace analog designers. Instead, these human designers will profit from the available tools which take over the laborious and routine tasks such that the designers can fully concentrate on new, innovative designs (of the first category) in which insight and expertise is required.

After this introduction of the terminology, the following section briefly reviews the characteristics of analog design and points out the similarities and differences with digital design, in order to give some indication for the complexity of the analog design automation task and to explain why analog design automation has been lagging behind its digital counterpart.

1.3. Characteristics of analog design

CAD and DA for digital VLSI has been a major research topic during the late seventies and the eighties. This has resulted in a general consensus on the different abstraction levels and the overall design strategy for digital system synthesis (though the solutions and algorithms for the subtasks can widely differ) [CAM_90, DEM_90]. Many methodologies, algorithms and programs have been presented [DEM_86, GAJ_88]. Many different design styles have been explored ranging from gate arrays over standard cells to silicon compilers. This research has also resulted in many commercially available tools, with a growing acceptability in the design community.

On the other hand, analog CAD and DA has not yet evolved that far. It is worldwide still in its first exploration phase. Analog design has for a long time been considered as too complicated and too unstructured to automate. The only tools available were numerical simulation programs, layout editors (polygon pushers) and filter synthesis programs (without automated layout). However, in recent years with the boom in the ASIC market, the automation of the total design path has become an economical necessity, also for analog integrated circuits [SAN_90]. This explains why, although the first pioneers began in the early eighties [ALL_83a, ALL_83b, DEG_83, DEG_84, THE_82], most of the analog CAD research started since the mid-eighties. As a result, the first prototype design programs are now being published [DEG_87c, ELT_89, GIE_90b, HAR_89b, KOH_90, etc.]. The first really commercial automated analog DA tool is the AutoLinear package of Silicon Compiler Systems (now Mentor Graphics) (which consists of the IDAC/ILAC programs of Marc DEGRAUWE [DEG_89]). It covers part of the analog design path, but it has only been released recently and it still has some serious limitations, as all other present tools. At the same time, several switched-capacitor filter compilers are now being presented [ASS_88, BAR_89, NEG_90, SIGG_87, THE_87, TRO_87, WAL_91]. Up till now, however, no general automated analog design systems have been presented yet which are also able to design higher-level modules such as A/D and D/A converters. The first prototype programs in this direction are presently being developed at research level [KEN_88, JUS_90]. Moreover, at present, there even is no general consensus among analog CAD developers on the different abstraction levels or on the overall analog design strategy to be used. This explains the wide diversity of the present solutions. The integration of the useful present ideas into a general analog DA concept will undoubtedly be a challenge for the near future.

To explain why analog CAD and DA tools are not so far developed as digital ones, one has to compare the characteristics of analog design to those of digital design. Analog and digital design differ in some critical points, making it impossible to take over existing digital CAD tools for analog without major modifications and adaptations. Nevertheless, all analog CAD and DA software should remain compatible with the digital tools in order to allow the automated design of mixed analog-digital systems.

The characteristics of analog design and the differences between the analog and digital domains are now briefly summarized below.

1.3.1. Loose form of hierarchy

A first difference relates to hierarchy. The hierarchical decomposition is clearly defined for digital systems with well-accepted levels (behavioral, functional, register transfer, logic, transistor) [CAM_90]. Analog designers on the other hand use only a loose form of hierarchy, as shown in the example decomposition of a Σ-Δ A/D converter in Fig. 1.7. The hierarchical levels are not so strictly defined and certainly not generally accepted. The same element can appear on several levels. Also, for the same function, one can easily interchange different-level components. For instance, gain can be achieved with one single transistor or with a total operational amplifier composed of many transistors. There is also a larger interaction between all levels. For instance, the frequency response of a circuit is important at all levels, but is determined by the interaction of all lower-level blocks down to the transistor level. Hence, many constraints, such as frequency or noise behavior, are passed down the hierarchy to the transistor level. The fundamental reason for this lack of any rigid, formal form of hierarchy for analog design is that the hierarchical decomposition in analog is based on a structural decomposition of the modules over several levels, and not on the properties of the signal types and the corresponding time representations at different abstraction levels as in digital. Instead, the variables in analog are voltages and currents and impedances at all hierarchical levels. This fundamental problem with analog will probably never be mastered. Nevertheless, it has been shown in [HAR_87] that hierarchy (in the meaning of structural decomposition) is also beneficial for analog integrated circuits. It allows to decompose the global, complicated design task in smaller, more manageable subtasks, and allows an easy reuse of existing design knowledge for each subblock. There is, however, not yet a generally accepted way of hierarchically decomposing an analog system. Also, to analyse or simulate higher-level circuits such as a Σ-Δ A/D converter, macromodels or behavioral models have to be used, since it is impossible to carry along all transistor-level details up the hierarchy.

1.3.2. Large spectrum of performance specifications

Secondly, many more performance specifications are imposed on analog circuits. For an operational amplifier, specifications include power, area, gain-bandwidth, DC gain, phase margin, gain margin, output range, common-mode input range, settling time, PSRR, CMRR, noise, input and output impedance, slew rate, offset, harmonic distortion, and so forth. A fairly complete list of specifications for an operational amplifier is provided in appendix A [SAN_88b]. In general, however, the set of characteristics for analog circuits is not easily fully defined, since some more specifications can always be added, such as temperature behavior, radiation hardness, etc. In addition, all these specifications can widely vary depending on the application. For example, the gain-bandwidth for a CMOS opamp can vary from 1 kHz (for biomedical applications) up to 100 MHz (for telecom and consumer applications with large oversampling). Also, for such large ranges and combinations of specification values, many circuit schematics exist which perform the same function, but each of which is most appropriate in a given range of specification values. In addition, the specifications often impose conflicting requirements on the design. This results in many trade-offs to be managed during the design of the circuit. This, however, is a multidimensional problem which is difficult to handle by a human designer. A typical example is the trade-off between bandwidth and gain, which also influences the noise and the area consumption. Human designers solve these problems by relying on their insight and previous experiences, and by using heuristics and rules of thumb.

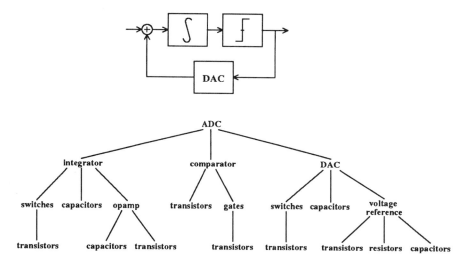

Fig. 1.7. Hierarchical decomposition for a Σ-Δ A/D converter.

1.3.3. Critical device sizing

Analog circuits typically have a smaller number of transistors per functional cell than digital circuits. However, there is a much stronger interaction between the electrical characteristics of each individual device and the performance of the global circuit [HAR_87]. For example, the phase margin of an operational amplifier depends on the capacitances and thus the sizes of each transistor. Hence, each transistor must be sized carefully, and the device sizes and component values can vary widely, depending on the specifications of the application. In practice, MOS transistor aspect ratios can differ up to a factor of 1000 [CAR_88c], or even go up to 10000 in some extremely-low-noise applications [CHA_90]. As a result, although analog circuits tend to have fewer transistors, the design time is nevertheless larger than for digital circuits because of this critical sizing. In digital circuits, transistors merely act as switches with minimum sizes. The size is only increased for transistors lying on critical delay paths or driving large loads.

1.3.4. Large range of circuit schematics

Digital circuits at logic or functional level are built up with regular cells, whereas in analog circuits not such a repetitive pattern of blocks is observed. Analog designers have developed a large number of circuit schematics performing the same function, such as for operational amplifiers, comparators, multipliers, etc. Each of these topologies has advantages and disadvantages and is most appropriate in a certain range of performance specifications. For instance, for lower gain values a single-stage opamp is sufficient, whereas for higher gain values a two-stage or multi-stage opamp is required; for high gain-bandwidths, a folded-cascode structure is appropriate, etc. Also, since analog circuits are moving towards high-performance applications (high frequencies, high precision, low noise, low distortion), more and more new topologies are being or will be invented to overcome the shortcomings of the present-day ones. In addition, the same topology can widely vary in sizes and component values depending on the specifications of each particular application. Hence, there are more diverse solutions for the synthesis of analog functions than there are for digital (at least at the gate or functional level, not at higher levels).

1.3.5. Large influence of technology

Analog circuits also show a larger influence of technology and environmental parameters. Process, biasing or temperature variations and layout parasitics strongly influence the circuit performance and can even change the functionality of the circuit. Hence, these effects are to be taken into account during the design. For example, the mismatch between devices mainly determines specifications, such as the offset voltage, the CMRR or the PSRR of a circuit. Changes in the substrate temperature strongly influence the frequency response of an operational amplifier. The matching precision obtainable in a given process is important since it can exclude certain circuit topologies for realizing a given design in that process. Similarly, analog circuits also require a precision modeling of the devices available in the technology. For example, for high-frequency opamps, the gate resistance of MOS transistors is important [OPT_90]. In addition, as opposed to digital layout, analog layout is not only concerned with minimum area. Instead, careful matching, symmetry conditions, routing parasitics, etc. are of more importance, since they can deteriorate the circuit's performance and turn a functional circuit (after the sizing) into a nonfunctional circuit (after the layout).

1.3.6. Interactions at the system level

Analog circuits are also very sensitive to interactions at the system level. A typical example is thermal feedback, in which an on-chip hot spot, say an output buffer, causes temperature gradients all over the chip. The change in local temperature can then influence the performance of some other temperature-sensitive stage [PET_90]. This performance perturbation is then propagated to the output and changes the power dissipation of the output buffer. This on its term changes the temperature gradients, and so forth. If there is some nonlinearity in this loop, it can even cause distortion. Similarly, if many different channels of a data-acquisition system are integrated on one chip, care has to be paid to avoid crosstalk between these channels [STE_87]. The problem is even worse when integrating both analog and digital functions on the same chip. Strong interactions are then possible through ' common power supply lines or through the common substrate. Since analog and digital circuits essentially behave in a different way, they also have a different sensitivity to these interactions. The most straightforward interaction is through the power supply voltage spikes at the switching times of the digital circuitry. If analog and digital circuits share the same supply line, these spikes can deteriorate the performance of the analog circuits.

1.3.7. High-performance applications

Finally, because of the increasing digitalization in integrated electronics nowadays, real analog design has been pushed to interface circuits for chips with a digital core and to applications with aggressive performance specifications, mostly at the edge of the limitations of the given technology. It is important to realize that the performance of analog circuits is only limited by physical phenomena. For example, the achievable bandwidths are limited by the fT of the transistors only, which is much higher than the maximum clock rate in digital circuits. The present analog circuits, however, are still far from their fundamental performance limits [TSI_87]. Hence, analog design nowadays is to a large extent associated with high-performance design, whereas digital is an appropriate solution for applications with not too aggressive performance specifications. This explains why analog circuits are used in applications requiring large dynamic ranges, high speed, high precision and linearity... [TSI_87, SAN_87]. Such applications of course require a careful design and layout of the analog circuits. To actually meet the high performances, often a full-custom implementation of the circuits is required, tuned to the particular application. The use of semi-custom methodologies, such as analog standard cells, is then excluded. Also, for these applications, performance-degrading second-order effects, such as parasitic couplings or distortion, become more and more important, since they limit the obtainable performances in practice [SAN_87]. Most of these effects, however, are still poorly understood by the analog designers and no heuristic rules exist yet to cope with these effects.

Conclusion

It can be concluded that analog design is a complex job to manage. It is commonly considered to be very knowledge-intensive and strongly relying on the insight and expertise of the designer, due to the complicated interactions between each individual device and the global circuit performance, between the circuit performance and the technology, and between different circuits on the same chip. Nevertheless, an analog designer succeeds in designing functional analog circuits by hierarchically decomposing the overall design task and by relying on his expertise and heuristics built up during many years. Obviously, **hierarchy and knowledge exploitation are major keywords for tackling the analog design automation task**. It takes however a long time for an analog designer to acquire his knowledge. The number of real analog experts is very limited. One can even question if real analog experts exist at all, since most analog designers are only experienced in a limited domain of analog circuits, such as operational amplifiers, oscillators or voltage references. Many companies also involve different people for the sizing and for the layout of a schematic. Summarizing, expert analog designers are hard to find and even harder to keep, whereas the training of novice designers is a long and expensive process [SAN_88a]. Besides, much design expertise leaves the company (or university) together with the designer. In order to avoid this, **one should structure, formalize and codify the existing analog design knowledge in a design framework and exploit this knowledge afterwards, in combination with powerful algorithms, for automated analog design or for the support of creative circuit designers**. The symbolic simulation techniques presented in this book also reduce the learning time of novice designers as well as increase the knowledge level of more experienced designers, by providing them with analytic insight into the behavior of analog circuits.

1.4. Needs for analog circuits and analog design automation

1.4.1. Application domains for analog circuits

The above overview of characteristics indicates that analog design is more complicated, more intuitive and far less systematized than digital design, and thus much more difficult to automate. This explains the present lack of mature analog design automation tools. On the other hand, the progress in the dense MOS VLSI technology, which now enters the submicron era, in combination with the commercial availability of advanced digital CAD tools, which allow to manage the vastly increased design complexity for the rapid and reliable synthesis of complex chips containing up to more than a million of transistors, has favored the digital implementation over the analog one for many functions and applications. As a result, during the eighties, integrated electronics have been characterized by an increasing digitalization, in which the analog part has been shrinking all the time and has been omitted wherever possible [CAR_88c]. Digital circuits, however, can never totally replace analog circuits since the signals in the outer world are analog. There is and will always be a need for analog circuits, even if it is just for interfacing. In addition, in recent years, a renewed industrial interest in analog circuits has been observed [TSI_87, COL_89], partially because of the tendency in ASIC design to integrate whole systems on one chip [SAN_90], partially because of many high-performance applications which are hard or even impossible to realize with digital circuits [CAR_88c].

These **application domains for analog integrated circuits** are now described in somewhat more detail :

• First of all, the signals in the real world (voice, music, electrical signals in the human body...) are analog in nature, with a continuous amplitude spectrum and time scale. Any electronic system which interacts with the outer world thus has to contain some analog *interface circuitry* (at least an analog-to-digital converter at the front end and a digital-to-analog converter at the back end). of such a system. Moreover, the specification requirements imposed on these analog or mixed analog-digital interface circuits are often very aggressive [OPT_90] and the performance of the total system can be degraded by the performance limitations inherent to these analog circuits. For example, the most critical part of an 18-bit oversampled D/A converter for audio applications (compact disc) is not the digital part, but the analog output filter. The noise and distortion generated by this filter limit the obtainable resolution of the overall system.

• Secondly, other noninterface functions can better be implemented with analog circuits as well. Some signal processing functions, such as multiplication or filtering, are easy to implement in bipolar, CMOS or BiCMOS circuits. Also, for many *high-performance applications*, analog can still outperform digital with less area, less power or in general a better overall performance [ALL_86a]. These application domains include :
   * *high-frequency applications* (extending into the GHz range for bipolar) such as television circuits, cellular mobile telephony, disc drivers. (Note also that the problems observed with digital circuits at higher frequencies are typically analog in nature.)
   * *low-noise data-acquisition systems* e.g. in biomedical sensor applications [STE_87], or in the front end of particle/radiation detectors for nuclear research [CAL_89] or space satellites.
   * very promising is also *parallel analog signal processing*, such as in neural nets where huge numbers of neurons (implemented as analog processing units) are to be integrated on one chip [MEA_89].

It can be concluded that analog circuits do have a great potential for high-performance applications, but that, due to the lack of analog CAD tools, designers instead in the past have been trying to avoid analog circuits as much as possible and to replace them by easier-to-design digital circuits [CAR_88c]. This provides a first motivation for the development of analog CAD and DA tools. **If analog design tools would be available, designers could fully exploit the analog capabilities** and benefit from analog instead of avoiding it. Analog could then be considered as a powerful alternative instead of an avoidable evil. At system level, it could then be investigated whether a digital or an analog implementation yields the best solution for any function for a given application and for the given state of the technology. The criteria for deciding between a digital or an analog implementation are then minimum power and area consumption, in general best overall performance, but also considerations such as testability, reliability, reproducability, programmability, etc. In this way, the most economical solution for a silicon implementation of an electronic system can be found while trading off all performances and costs.

1.4.2. Industrial need for analog design automation

In addition to these high-performance applications, there nowadays is also a large industrial demand for analog DA tools in order to reduce the design time of mixed analog-digital ASICs. In recent years, there is a pervasive trend towards the integration of complete electronic systems, which before took a whole printed circuit board, into one single silicon chip, especially for ASICs. This implies the integration of both the digital and the analog (interface and noninterface) circuits on the same silicon. One speaks of **marrying digital and analog functions on chip to achieve high performances** [SAN_90]. These integrated systems are also called **microsystems**. Typical examples are telecommunication circuits (such as a modem chip), audio circuits (for instance for compact disc), other consumer electronics (such as many electronic toys), automotive circuits, intelligent sensors, etc. Today, about 25% of all ASICs are mixed analog-digital, and this percentage is predicted to increase rapidly. The reasons for this increasing integration are economical, such as reduced cost, reduced power consumption and increased reliability. For consumer electronics, the main driving force to develop ASICs is a reduction in the production costs, which is only possible because of the large production volumes since the initially larger design effort and cost have to be recuperated.

As already explained before, due to the lack of mature analog CAD tools and problems with analog testability, reliability, reproducability, etc., many functions in these ASICs are nowadays implemented with digital circuitry, which can be synthesized rapidly and reliably by means of advanced digital CAD tools. The few analog circuits, however, are still mainly being handcrafted by expert designers and thus require large design times. They also cause most of the design errors in present ASICs, resulting in expensive reruns. **Handcrafted and likely error-prone analog design cannot compete with automated and highly structured digital design for most applications!** In addition, the integration of both digital and analog circuits on the same chip causes additional problems, such as unwanted interactions which can strongly influence the performance of the analog circuits and hence the functionality of the overall system. As a result, whereas the present ASICs and VLSICs consist of a large digital part and a small analog part, **the small analog part still takes an inversely large part of the total design time**. This is qualitatively shown in Fig. 1.8, where the fraction of the area occupied by digital and analog circuits is opposed to the fraction of the design time required for designing these digital and analog circuits. Typically, the analog circuits occupy only 10% of the total area, but consume 90% of the total design time and cost.

VLSICs, ASICs

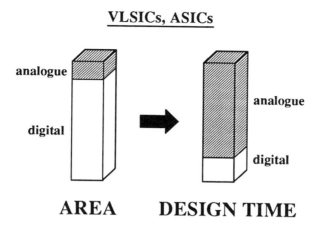

AREA   DESIGN TIME

Fig. 1.8. Relative area occupation and design time consumption for analog and digital circuits in present VLSICs and ASICs.

Hence, the design of the analog parts can be the bottleneck in the design of many mixed analog-digital systems. This is also critical from an economical point of view. The time to market nowadays is crucial for most consumer electronics in the face of rapidly evolving technology and decreasing product life cycles. A rerun caused by a failure in the analog circuitry can result in a late product introduction, a reduced market share and a great loss of profit. Hence, analog computer-aided design tools and analog design automation tools are badly needed. This is the second motivation for the development of analog CAD and DA tools.

Summarizing, the pressure from the ASIC design industry and the need for analog circuits in many high-performance applications explain the renewed interest in analog circuits and the large industrial demand for analog CAD and DA tools observed today. This book provides a brief introduction into this new field and describes the major approaches presently being explored. Note however that we are still far from commercially available total analog design automation solutions.

1.4.3. Objectives and requirements for analog design automation

If we summarize all the above considerations, then the **goals of analog CAD and DA tools** are the following [GIE_90a] :
• shorten the overall design time from concept and behavioral system description down to silicon. The resulting silicon must be guaranteed to work correctly from the first time (*correct by construction*) and it has to be testable.
• formalize and capture the existing knowledge of analog designers. This knowledge has to be transferred to the CAD system and exploited afterwards.

- provide insight to the tool user. This allows the transfer of the built-in knowledge to novice designers and shortens their education time.
- assist circuit designers by taking over the routine tasks, such that the designers can concentrate on the creative work.
- make analog design available for system designers by automatically designing analog circuits from system specifications.

In order to realize these goals, the analog CAD and DA tools are to cover the entire spectrum of analog integrated circuit design over multiple abstraction levels and are to be integrated into one framework with a common database, compatible with similar digital tools. The **requirements for such an analog design automation system** are the following.

- Because the performance and design time requirements can largely vary from application to application, different design styles can be used for designing the ASICs and VLSICs. These design styles will be described in the next section. However, the DA system should be able to support these different styles. For example, for critical high-performance applications, the circuits really have to be optimized in terms of area, power and overall performance, and thus have to be implemented in a full-custom way. For less critical ASIC designs or for lower-volume applications, other approaches such as linear arrays or standard cells can be applied as well. At the same time, the DA system should be independent of any particular process used and allow both CMOS, bipolar and BiCMOS technologies (and perhaps other technologies in the future).
- Because the performance specifications can vary largely and because each circuit schematic is only appropriate in a limited performance range, the system may not be limited to a fixed set of topologies. Instead, it should be able to easily introduce new circuits into the DA system by the designer himself [GIE_90a, GIE_90b].
- To allow the design of complex systems, the complex design task at top level has to be divided into more manageable, smaller subtasks i.e. the DA system has to be hierarchically structured over several levels of design abstraction [HAR_89b] with the appropriate description languages at each level and the ability to automatically translate specifications downwards and detailed design information upwards in this hierarchy.
- Analog design is very knowledge-intensive due to the many interactions and trade-offs to be managed during the design. Human designers only succeed in designing fairly good analog circuits by relying on their insight and expertise. This expert knowledge consists both of formal (analytic) knowledge and intuitive (heuristic) knowledge [ELT_89]. Hence, the DA system should capture this knowledge and exploit it for all designs later on [CAR_88c]. However, care must be taken to the way how the knowledge is extracted out of the expert designers. Analog experts cannot be expected to program the way they design a circuit. First, they often are not capable to describe in a formal and general way how they perform a particular design. **One of the largest problems in analog design actually is this lack of structured analog design knowledge.** Secondly, designers are mostly

inexperienced with programming techniques and are certainly not familiar with the implementation details of the DA system. Hence, a designer-friendly interface format should be available for the knowledge acquisition. In the most ideal case, self-learning capabilities will allow the system to learn from the designers itself.

• The system has to address **two kinds of users: system designers and circuit designers.**

  * *System designers* design at the system level and usually are unfamiliar with all circuit implementation details. They only want the DA system to automatically design the circuits and modules they need at the system level, in a very short time and such that these circuits meet all specifications imposed from the system level. These designers want to run the system in automatic mode, relying on the built-in knowledge.

  * *Circuit designers* design analog modules and circuits down to the transistor level relying on their own experiences. Their design tasks can be divided into routine tasks (which are not innovative and require more transpiration) and nonroutine tasks (which are innovative and require more inspiration). The CAD system has to free these circuit designers from the routine tasks, such as the manual calculation of analytic formulas or the generation of the layout for noncritical applications. In this way, the designers can fully concentrate on the creative part of the design, such as designing circuits for aggressive specifications or exploring new circuit schematics. For them, the CAD system is more a design assistant which takes over or supports the laborious routine tasks. The circuit designer himself will always take the ultimate design decisions. Hence, for the CAD system to be acceptable for these expert designers, the system really needs to be open. The designers always have to be able to put in new knowledge and to ask explanation for any decisions taken by the system [GIE_90a].

After this summary of the objectives and requirements for analog design automation, the following section briefly reviews the different approaches or design styles presently used in analog design, resulting in the discussion of analog silicon compilation.

1.5. Different analog design approaches and analog silicon compilation

1.5.1. Overview of different analog design approaches

In the past, several approaches have been used for the design of analog integrated circuits [ALL_86a]. Most of these techniques resemble previous digital approaches. All styles can implement the same analog circuit or system, but differ in production cost, turnaround time, flexibility and achievable performance depending on the degree of customization. Hence, the selection of the most appropriate design approach for a particular application depends on criteria such as the production volume and the requested performance level.

The **different analog design styles** are now categorized below according to increasing customization :

- *programmable chips*
  Programmable chips are completely prefabricated chips which can be programmed through electrical (e.g. RAM programmable) or other means (e.g. fuses) [COX_80]. They have a zero turnaround time, but the most limited performance and flexibility and the smallest application range.

- *linear arrays*
  Transistor arrays consist of large prefabricated arrays of unconnected fixed-size transistors, resistors and capacitors [KAM_83, KEL_85, KIN_85, PIC_83]. The interconnections between the devices still have to be programmed by the user. Such transistor arrays have a very fast turnaround time since only the interconnection layers have to be processed. However, although the performance can somehow be modified by combining transistors in parallel or in series, the occupied area is large and the overall performance is still low. For example, due to the predefined placement and the large routing, offset and parasitic couplings are difficult to control. If these transistor arrays also contain some predefined but externally unconnected building blocks, such as switches, or total circuits, such as opamps, they are called linear or circuit arrays. These arrays are similar to gate arrays for digital design. Also, recently, analog sea-of-gates approaches were announced which combine fair densities and performances with moderate turnaround times.

- *analog standard cells*
  Analog library cells are fixed, predefined cells which have been laid out by experts. If they all have the same height, they are called analog standard cells [KEL_85, PLE_86, SMI_89a, STO_84]. These cells have to be selected, placed and routed for each application. The turnaround time is moderate. All cells have a proven quality, but the design flexibility is limited since only a discrete set of cells are available, each useful for a limited range of performance specifications. Moreover, these standard cells become obsolete and have to be redesigned whenever the technology process changes.

- *parameterizable cells*
  Parameterizable cells (also called module generators) are predefined cells, where part of the circuit topology or the layout is fixed and the user is able to parameterize the remainder. Usually the circuit topology is fixed and the user can vary the transistor sizes or passive component values [ALL_85, KUH_87, STO_84, WIN_87]. Actually, such a module generator is nothing more than a procedure which generates the cell for the parameters entered by the user. Often, this is implemented under the form of a parameterized layout with a fixed floorplan and sizable or stretchable devices. This approach allows a moderate turnaround time. In practice, however, the parameterization and hence the

achievable performance range usually are limited. Moreover, each module generator is restricted to one particular cell or block only. As a result, this method is mainly useful for regular structures, such as switched-capacitor filters [YAG_86]. Also, the analog module generators in the CONCORDE silicon compiler system are based on parameterized layout for common analog functions [KUH_87].

• *full-custom design*

In full-custom design, the selection of the circuit topology, the sizing of all devices and the generation of the layout are performed for each application starting from the given performance specifications and the technology data. It has the largest flexibility and achievable performance with the lowest area. However, the design takes much more time and the total turnaround time is large. Economically, full-custom design is only affordable for large production volumes.

The characteristics of these different analog design styles are now summarized in table 1.1. Note also that each of these analog design styles can be automated. Since each approach has a different design complexity, the complexity of the corresponding design automation task also widely differs. For transistor arrays only the interconnections between the fixed devices have to be generated automatically, whereas for full-custom design both the topology selection, the sizing and the layout have to be implemented. As shown in table 1.1, programmable chips, linear arrays or analog standard cells have a relatively small turnaround time, at the cost of limited flexibility and achievable performance. Analog module generators have more flexibility to tune the cell to each application, but the parameterization is usually limited as well and each module generator is dedicated to one cell only. Hence, all these semi-custom methods provide a rapid path to silicon, but they suffer from area/performance limitations which cannot be accepted for many high-performance applications [HAR_89a]. The semi-custom methods are thus appropriate for many moderate-performance designs, but for high-performance applications, the circuit has to be designed and tailored to each application in a full-custom way.

	density	flexibility	performance	turnaround time
progr. chips	--	--	+/-	+++
analog arrays	-	+	+/-	+
standard cells	+	-	+	-
param. cells	+	+	+	-
full custom	++	++	++	---

Table 1.1. Comparison of density, flexibility, performance and turnaround time for different analog design styles.

To give an idea of the relative importance of the different design styles on the market, Fig. 1.9 shows the present distribution over the different design styles for the commercially available ASICs. Remember that about 25% of all ASICs are mixed analog-digital nowadays. About half of the ASICs are implemented on gate arrays (eventually with some analog subarray) or linear arrays. About 25% of the ASICs are realized with standard cells. About 15% are designed in a full-custom way, and the remaining 10% are implemented with programmable chips, such as PLAs, etc.

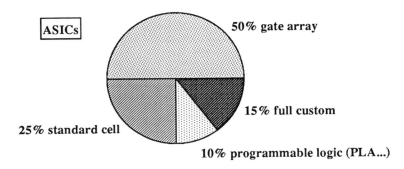

Fig. 1.9. Distribution over the different design styles for the present commercially available ASICs.

From this discussion, it is clear that all design styles have a different market segment, and that a general DA system should include different styles. However, because analog circuits nowadays are largely associated with high-performance applications, there is a large industrial demand for tools to automatically design analog circuits in a full-custom-like way [DEG_87c, GIE_90b, HAR_89b, KOH_90, etc]. Besides, the semi-custom methods are more oriented towards laying out an already-designed circuit (e.g. place and route tools for standard cells). They do not provide any means for really synthesizing custom analog circuits starting from specifications and technology data. For this, the concept of analog silicon compilation and analog module generation are introduced.

1.5.2. Analog silicon compilation

First, the definition of an analog silicon compiler is presented.

Definition: analog silicon compiler

A silicon compiler is a computer-aided design system which automatically translates a high-level system description (behavioral specification) into a physical layout (silicon) that performs the specified behavior (while possibly also optimizing some cost function, such as minimum power or area). At the input of the compiler, the behavior of the high-level system under design is described by means of some behavioral language, together with a set of performance and environmental specifications. (There is no general agreement yet on the analog behavioral description language to be used.) The output of the compiler is then the layout (mask patterns) used for the silicon implementation of the system. The resulting silicon after processing then has to be guaranteed to satisfy all system performance specifications. In this way, **an analog silicon compiler automates the analog design path for analog (sub)systems from the system's behavioral description down to layout** over several levels of hierarchy.

A similar definition can be provided for an analog module generator.

Definition: analog module generator
An analog module generator automates the analog design path for analog functional modules (building blocks) from the module's behavioral description down to layout. Given these definitions, all existing analog design systems [DEG_89, GIE_90a, HAR_89b, KOH_90, JUS_90, etc.] are analog module generators and not analog silicon compilers. The latter, however, is the ultimate goal of our research.

What are the **advantages of silicon compilation**? Some of the reasons for using an analog silicon compiler are :
• it allows high-level system design
• first-time correct design (correct by construction)
• fast design with low design cost
• application-specific power/area/performance optimization possible
• automatic inclusion of testability and yield considerations
• easy design update when technology changes

Also, practical analog compilers or module generators will most likely be limited to the automated design of certain restricted classes of analog architectures (e.g. for signal processing) or modules (e.g. operational amplifiers, switched-capacitor filters, or A/D and D/A converters), within some performance limits (not too agressive specifications). To this end, they exploit the widely available knowledge built up for these architectures or modules during many years. However, if carefully conceived, the framework for implementing for example the analog module generator for a certain class of analog modules may be reused for several other classes of analog circuits. In addition, the compiler or module generator may use any of the design styles discussed in the previous subsection, although for most high-performance applications, a fully customized implementation of the analog circuits, optimized and tailored to the actual application, is required. In the remainder of this text, we will now focus only on analog design systems supporting such a full-custom design style.

The **key features for an analog silicon compiler or an analog module generator** can then be summarized as :
- hierarchically structured
- exploit expert domain knowledge
- allow multiple topologies and easy extendability
- application-specific optimization
- combining automatic design mode and user explaining facility
- compatibility with digital CAD tools and compilers

It is also important to realize that such an automated analog design system will always be limited to designs with not too agressive performance specifications. Designs with real aggressive demands still have to be carried out by expert designers, which can run parts of the system to take over the laborious, routine tasks while they can themselves concentrate on the more creative parts of the design. On the other hand, designs with moderate to reasonably high specifications can be carried out fully automatically by an analog silicon compiler or module generator.

After this discussion of the different analog design styles and the features and possibilities of analog silicon compilation and analog module generation, a design strategy for the automatic synthesis of analog or mixed analog-digital integrated systems is now described in the next section. According to this strategy, a major distinction between system synthesis and module synthesis is made.

1.6. Analog system-level synthesis

Applications for telecommunications, audio, video, biomedicines, instrumentation, particle/radiation detection, sensor interfaces, automotive, etc. require the design of analog and mixed analog-digital systems. All these systems have to transform input signals into output signals according to some behavioral description. Consider for example a biomedical data-acquisition chip, which is part of a general implantable Internal Human Conditioning System (IHCS) [EYC_89]. This IHCS, shown in Fig. 1.10, monitors signals in the human body (e.g. the glucose concentration in the blood) and eventually takes some corrective actions (e.g. injection of some extra glucose by switching on a small implanted pump). The data-acquisition chip is the interface between the sensor electrodes and the microprocessor, which forms the core of the system and which interprets the sensor signals and decides on the corrective actions. Hence, the functions to be performed by the data-acquisition system are mainly signal conditioning and signal processing (gain and filtering) before the A/D conversion. The behavioral description of this system at top level can then be stated as: it has to convert weak and noisy analog sensor signals into strong digital signals for a given number of channels. Additional specification values are the signal frequency range, the required dynamic range (limited by noise and distortion), etc.

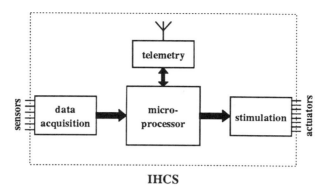

Fig. 1.10. Block diagram of the internal human conditioning system (IHCS).

A suitable design methodology for such analog and mixed analog-digital systems is then based on the **meet-in-the-middle design strategy** [DEM_86]. The overall design process is split up in a system design part and a module design part. At the system level, the system is decomposed into an interconnection of analog building blocks or modules (such as a filter, a modulator or an output buffer), which are then automatically designed by an analog module generator according to the translated specification values. This decomposition substantially eases the design task. It allows a clear distinction between system design and module (circuit) design and hence between system designers and circuit designers. In this way, system designers can concentrate on system synthesis, architectural issues, system-level trade-offs without worrying about all circuit implementation details. On the other hand, circuit designers can concentrate on designing the modules in the case of aggressive performance specifications given by the system designers. For more relaxed specifications, the design is performed automatically by the analog module generator.

An analog silicon compiler implementing the above design strategy has, at the system level, to translate the behavioral description into an interconnection of analog functional modules with appropriate module specifications drawn from the system specifications (the modules are then automatically designed by the module generator part of the compiler), to assemble the returned layouts and to verify the resulting system design. However, up till now, no real and mature tools have been presented at the system level yet. The research field of analog high-level synthesis is yet unexplored thus far.

A block diagram of the different tasks to be performed at the system level is shown in Fig. 1.11. The CAD tools required at this level are summarized below.

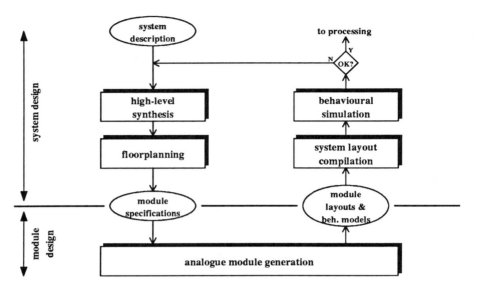

Fig. 1.11. Block diagram of analog system-level design tasks.

1.6.1. Analog high-level synthesis

At the system level, only the behavioral description is given of what the analog or mixed analog-digital system has to achieve and within which performance specifications. The analog high-level synthesis program then automatically has to synthesize the system by partitioning it in (relatively noninteracting) analog functional modules or building blocks. This first requires the selection of the analog modules needed to realize the requested system behavior (selecting or setting up the system's *architecture*) and then the determination of the specifications for each module from the specifications for the system (*translating or mapping the specifications* downwards).

A possible architecture for the data-acquisition chip of Fig. 1.10 is shown in Fig. 1.12 [STE_87]. It consists of an instrumentation amplifier, a gain stage, an anti-aliasing filter, a programmable filter, a sample-and-hold, a multiplexer and an A/D converter. This decomposition into an interconnection of analog functional modules starting from the system's behavioral description is the first task of analog high-level synthesis. The next step is the translation of the system's specifications into specifications for the individual modules in the system's block diagram. This means determining the specifications for the instrumentation amplifier, the filter, etc. in Fig. 1.12 from the specifications for the whole data-acquisition chip. This translation is subjected to the following condition: it has to be guaranteed that if all modules satisfy their translated performance specifications, the total system also meets its performance specifications. To guarantee this, behavioral simulation can be used (see subsection 1.6.3 below). At the same time, a floorplanner keeps track of the overall organisation of the modules on the system's floorplan. This geometry information is passed down to the individual modules as well.

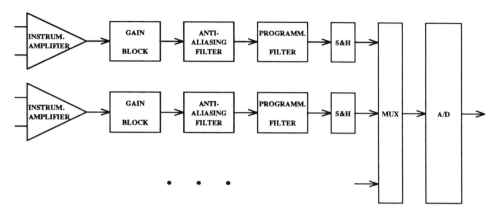

Fig. 1.12. Multi-channel architecture for the data-acquisition chip in the IHCS system.

The modules are then designed independently by the module generator part of the silicon compiler. This module generator optimizes every building block according to the specifications imposed from the system level and then automatically generates the layout. A flexible methodology for the automated design of analog functional modules, together with other approaches presented in the literature, will be described in the next chapter.

There is also an additional degree of freedom in system-level design. Each of the functional modules can be implemented with digital or with analog circuits. The selection between an analog or a digital implementation for a given module with given specifications and for a given state of the technology is thus a trade-off to be managed at the system level, based on criteria such as power and area consumption, testability, reliability, reproducability, etc. An example of such trade-offs can be found in the data-acquisition chip of Fig. 1.12. In the architecture shown there, the filters are analog and the A/D converter is at the back end of the chip. However, one can as well place the A/D converter somewhere in the middle and then perform the filtering with digital circuits.

Summarizing, the high-level synthesis system has to determine which modules are needed, how they are interconnected and which specifications each block must meet. This is a complicated, nontrivial task which has no unique solution, also because of the many system-level trade-offs to be managed. Up till now, no one has addressed this problem. Most likely, it requires a combination of knowledge-based techniques with formal circuit behavior theory. However, it can be questioned whether fully automated design of analog or mixed analog-digital systems will ever exist. Perhaps it is feasible for some predefined system architectures, for example in signal processing applications. In general though, a more realistic approach will be the development of performance-driven expert tools to aid system designers with their system-level synthesis. This is still a challenging research area for the near future. On the other hand, tools for automating the design of analog functional modules have evolved much further, as will be described in the next chapter.

1.6.2. System-level layout compilation

After the analog modules are designed by the module generator, the module layouts are passed upwards to the system level. The layout of the total system is then generated by assembling the layouts of the different modules. This occurs by placement and routing tools, which take the many analog layout constraints into account. For example, for the placement, requirements resulting from thermal considerations become important. For the routing, an efficient mixed analog-digital router is needed which deals with constraints, such as the minimization of the parasitic capacitive and resistive load on sensitive nodes, the avoidance of critical couplings, etc. Such tools are currently under development at several places [CHO_90a, CHO_90b, GYU_89, MAL_90, PIG_90].

1.6.3. Analog behavioral modeling and system-level simulation

After the system layout has been compiled, a functional verification of the entire, possibly mixed analog-digital system has to be performed. With a numerical circuit simulator such as SPICE [VLA_80b], this would require giant CPU times because of the combination of large simulation periods with small time steps needed to simulate the fast transients, which makes the use of circuit-level simulators impossible in practice. Instead, behavioral simulation is needed in which all circuits are replaced either by an appropriate behavioral model or by a macromodel. This implies the development of an efficient mixed-level mixed-analog-digital system simulator, of a behavioral modeling language which is suited for describing analog functions, and of an automated modeling tool to generate the appropriate models. A possible candidate for the analog behavioral language is the Analog Hardware Description Language AHDL presented in [KUR_90]. Later on this language has to be merged with a digital behavioral language, such as VHDL [VHD_87]. The behavioral simulator then has to perform a functional verification of the entire system, which also includes second-order performance-degrading effects, such as temperature feedback or power supply couplings. Hence, these parasitic effects have to be included in the behavioral models for the analog modules in addition to the nominal behavior. If the simulated performance deviates from the specifications, then the results of the simulation are fed back to the system-level synthesis program. The behavioral models themselves have to be generated automatically by the module generator after completing the design of each module, starting from the extracted layout. These models are then passed upwards to the system level together with the layout of the module. Much research is going on in this field of behavioral modeling and simulation [COT_90, KUR_90] and some simulators have already been presented. For example, SABER is a mixed-mode mixed-level simulator [SAB_90] which is becoming very popular nowadays. Another approach based on piecewise linear approximations is implemented in PLATO [STI_90]. At this moment, however, there is not yet any general consensus about the simulation algorithm, the behavioral description language, the effects to be included in the behavioral models for analog circuits, etc.

1.6.4. Analog testing

Finally, an extremely important but terribly difficult task is the testing of mixed analog-digital systems. The problem can more or less be tackled adequately for digital circuits with scan-path and built-in self-test techniques. But for analog circuits, the problem is far from being solved. This arises from the fact that the quantities of interest during the testing are continuous, sensitive waveforms, which may not be effected by the additional test circuitry, and that most failures are parametric in nature (i.e. is the system within the specifications or not). Some work has already been presented in this area [MIL_90]. Much more research, however, still has to be carried out before a general strategy can be proposed to make analog designs testable and before an analog silicon compiler can automatically include the required test structures on chip and generate the corresponding test programs. This is a challenging and important research topic for the near future. At the same time, many analog designs are presently being completed without care of the testability problem. The testing afterwards is simply performed by applying large quantities of test vectors and waveforms. Some of these tests may even be redundant. Therefore, analog circuit designers may have to reorient themselves and start designing with testability in mind.

Conclusion

In this section, the overall design process for mixed analog-digital integrated systems has been split up in a system design part and a module design part. The design tasks at the system level have briefly been summarized as high-level synthesis, layout compilation, behavioral simulation and analog testing. The research activities in these fields are still in the initial exploration phase. On the other hand, for the automatic generation of the analog functional modules, the research has much more evolved and has resulted in several practical programs. This will be the topic of the remaining chapters.

1.7. Outline of the book

In this first chapter, we have described the overall context of this book, the automated design of analog and mixed analog-digital integrated circuits and systems, by providing a general introduction to analog design automation, by explaining the terminology of the domain and by indicating the industrial motivations for this research.

The need for analog circuits (both for interface circuits to the analog world and for high-performance applications) and the economics-driven tendency in ASIC and VLSIC design to integrate total mixed analog-digital systems onto one chip today result in a large industrial demand for analog design automation tools. These analog tools, however, have been lagging far behind their digital counterparts because of the complicated and less systematic nature of analog design. Therefore, an analog silicon compiler is needed which automates the analog design process from system specifications down to layout and which - for high-performance applications - tailors the design to each application in order to obtain an optimal performance at the lowest cost (area and power).

The overall design task for analog and mixed analog-digital systems has then been divided in a system design part and a module design part. Up till now, no system-level design tools are available yet. Chapter 2 therefore concentrates on the automated design of the analog functional modules (such as operational amplifiers, comparators, filters or converters), for which several successful programs have been presented. The basic concepts are introduced and illustrated by the discussion of a particular methodology [GIE_90a, SWI_90]. The key aspect of this methodology, which tailors each circuit to the specifications of the individual application, is that it uses symbolic simulation and design equation manipulation as a solution to build an open analog synthesis system, in which the inclusion of new circuit schematics and design knowledge is highly simplified. This then provides us with the motivation for the detailed discussion of symbolic simulation, the major topic of the book, in chapters 3 to 5. The methodology is also compared to other approaches to properly depict the state of the art in this field.

In chapter 3, the functional aspects of a symbolic analysis program for analog integrated circuits is presented [GIE_89c]. This program, called ISAAC, derives analytic expressions for all AC characteristics of a circuit as a function of the complex frequency variable and the symbolic circuit parameters. It also provides a heuristic approximation of the symbolic expressions in order to retain the dominant terms only. The program is used to automatically generate the analytic models needed in the analog module design methodology of chapter 2 and to provide insight into a circuit's behavior to both novice and experienced analog designers.

The algorithmic aspects of the program are then described in great detail in chapter 4. These include the set-up, the solution and the approximation of the circuit equations. This results in the discussion of the overall performance and efficiency of the program and a detailed comparison to other techniques and programs. It also clearly points out the possibilities and the limitations of symbolic simulation as for the time being.

In chapter 5, the basic techniques for linear circuits are then extended towards the symbolic analysis of the harmonic distortion in weakly nonlinear analog circuits. The chapter also briefly discusses other recent developments in the field of symbolic simulation, such as hierarchical attempts for the simulation of large circuits. This then clearly depicts the ongoing activities and the state of the art in this field, which is of growing interest to the designers.

Finally, chapter 6 shows how these symbolic analysis techniques are used in the design system presented in chapter 2 to put new knowledge into the synthesis system and to apply this knowledge to size practical circuits. The method, which is based on optimization [GIE_90b], is illustrated with a practical design problem. After the sizing, the design is then completed by the layout generation, which is only discussed briefly.

2   THE AUTOMATED DESIGN OF ANALOG FUNCTIONAL MODULES

2.1. Introduction

The design of integrated systems, including both digital and analog parts, has in the previous chapter been divided in a high-level system design part and a module design part. At the system level, the overall system is then decomposed into an interconnection of functional modules, such as A/D and D/A converters, buffers, filters, oscillators, etc. Each of these modules performs a well-defined function at a certain hierarchical level and can consist of several submodules from a lower level. For example, a filter can include an integrator, an integrator comprises an operational amplifier, an operational amplifier contains a current mirror, a current mirror finally consists of transistors. The analog modules are then automatically designed by an analog module generator according to the specifications imposed from the system level.

For this, several different approaches and tools are currently being published. Note that in this text we focus on systems which design the analog circuits in a full-custom-like way for each application. A classification of the different design programs is then presented in the next section 2.2. The classification is based on the techniques used for the basic tasks within every analog module design program: topology generation, sizing or dimension generation, and layout generation. In addition, the design systems can also generally be divided in two classes: open and closed systems. These concepts are then illustrated in section 2.3 by the presentation of a flexible methodology for the automated design of analog functional modules. The major feature of this methodology is that it uses symbolic simulation and equation manipulation techniques as a solution towards an open design system. Indeed, to carry out the design, each module is characterized by an analytic model, the AC characteristics of which are generated automatically by means of a symbolic simulator. As a result, the inclusion of new topologies into the design system is largely simplified and the design system is not limited to a fixed set of topologies. This then clearly provides us with the motivation for the detailed discussion of symbolic simulation techniques in the following chapters 3 to 5.

The automated analog design methodology, which covers the whole design path from topology generation over analytic modeling and optimal circuit sizing down to layout, is then illustrated for the design of an operational amplifier in section 2.4. Finally, section 2.5 also briefly discusses the most viable and most matured other analog design systems presented in the literature up till now. This then results in a critical comparison of all methods, discussing the advantages and limitations of each design system and clearly depicting the state of the art in analog module generation at this moment.

2.2. Classification of analog module design programs

For the automatic synthesis of analog building blocks, several approaches and tools have been published [ALL_83ab,84,85,86ab, ASS_87,88, BAR_89, BER_88ab,89, BOW_85,88,89, BRA_89, CAR_88abc,89,90, CHE_89, CHO_90ab, CHR_86, COH_90, CRU_90, DEG_83,84,86,87abc,88ab,89,90, DEK_84b, DON_90, ELT_86,87,89, FUN_88, GAR_88, GAT_89, GEA_89, GIE_86,89abc,90abcd, GOF_89ab, GYU_89, HAB_87, HAR_87,88ab,89ab, HAS_89b, HEI_88,89, HEN_89, HSU_89, IPS_89, ISM_90, JUS_90, KAW_88, KAY_88ab,89, KEL_85,88,89, KEN_88, KIM_89, KOH_87,88,89,90, KON_88,89, KUH_87, LAI_88, LEE_90, MAI_89, MAL_90, MEI_88ab, NEG_90, NYE_81,88, OCH_89, ONO_89,90, PIG_89,90, POR_89, RAA_88, RES_84, RIJ_88,89, ROY_80, SAN_87,88a,89a, SAS_89, SCH_88, SED_88, SHE_88, SHY_88, SIG_87, SMI_89bc, STO_86,88ab, SWI_89,90, THE_82,87, TOU_89, TRO_87,89ab, VER_88, WAL_87,89,91, WAM_90, WAW_89, WEB_89, WIN_87, YAG_86]. In each of these programs, the analog circuits are resized for every application, to satisfy the performance constraints, possibly also optimizing some objective function. The programs all use some analytic description (design equations) of the circuit to perform the sizing and all exploit expert designer knowledge to some extent. A possible classification of these programs is now discussed below.

2.2.1. Open and closed design systems

In the area of analog DA, two trends are now emerging, resulting in **two different classes of analog design systems** [CAR_90] :

- *closed systems* :
   these are unchangeable systems in which only the circuit topologies and the design knowledge added by the tool creator are available to the designer [DEG_87c; ELT_89, HAR_89b, KOH_90].
- *open systems* :
   these are systems that can be extended by the designer himself [GIE_90a, OCH_89]. They can be described as open environments in which the designer can modify the existing design expertise, as well as easily add his own knowledge and new circuit schematics, which are then stored for reuse later on. Therefore, these systems are often also called **non-fixed-topology systems**.

In a closed design system, the designer cannot modify or extend the system by including new schematics or knowledge. He often even cannot inspect the system's knowledge base, nor does he get information about the many heuristic decisions and simplifications introduced by the program. In addition, as general knowledge hardly exists, it is often difficult, if not impossible, to use the program outside its implicitly-defined target application domain. This explains why analog circuit designers are resistant to such closed, opaque design systems. They prefer open systems in which they can access and eventually modify all internal knowledge, in which they are informed about all design decisions and simplifications introduced by the design system and in which they can include their own knowledge and new circuit schematics as well. Such **open design systems are therefore desirable for circuit designers**, who are often confronted with aggressive specifications and challenging designs, which cannot be solved with the existing design knowledge. The main problem with these open systems, however, is how to make the process of adding design knowledge easy [CAR_90]. A solution for this adopted in this book is the use of symbolic simulation.

On the other hand, **for system designers, who want as less interaction at the circuit level as possible, a closed design system can be sufficient as well** for many applications, as all design knowledge about some well-understood analog topologies can be encapsulated in the program and can then be used to automatically generate acceptable design solutions in a short time. It can be concluded that open and closed systems are targeted to different groups of designers. Note, however, that many open systems, such as the one presented in this chapter [GIE_90a], can be used automatically as well, relying on the built-in knowledge, and thus actually address both groups of users. In addition, the accumulated design expertise introduced in an open system by expert designers will later also become available for automated design to system designers.

The design of analog modules in any of these design systems consists of three basic tasks, as already explained in section 1.2 in the previous chapter: topology generation, sizing (or dimension generation) and layout generation. Depending on the techniques used for each of these design tasks, all design systems and programs can now be divided in different categories.

2.2.2. Classification based on the topology generation method

The first task is the generation of an appropriate circuit topology for the given performance specifications and technology data. For this task, the following methods can be distinguished :

* *top-down topology selection :*
  in these methods a topology is selected from a limited number of fixed, predefined alternatives, which are stored in the design system's database [DON_90, HAR_89b, KOH_90]. The topologies in these **fixed-topology systems** can be flat transistor schematics or can be built up hierarchically in terms of lower-level

blocks. In the latter case, the selection has to be performed at each hierarchical level, while moving down in the hierarchy. The selection between different topologies is carried out by means of heuristic selection criteria, which take the actual specification values and the technology data as input parameters. This selection process can then be represented in a decision tree. An example of such a tree for the selection of an operational amplifier is shown in Fig. 2.1.

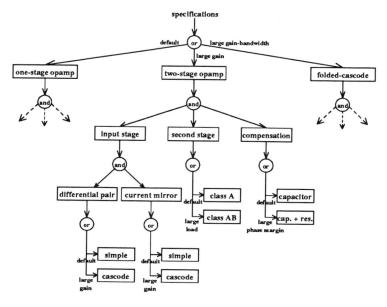

Fig. 2.1. Example decision tree for the selection of an operational amplifier topology.

If more candidate topologies are viable, then there are two possibilities [HAR_89a] :

* all viable candidate topologies can be ranked, for example by means of some heuristics estimating the probable area and power consumed by each topology for the given specifications. Then the first topology in this ranked list is synthesized. If it fails to meet the specifications, the second topology is synthesized, and so on. The first topology which meets all specifications is accepted as final solution. The advantage of this approach is that with proper selection and ranking heuristics only one or in the worst case a few topologies have to be designed.

* the alternative is to synthesize all viable candidate topologies and finally pick out the best one, for example the one with the smallest area occupation and/or power consumption. This approach is more expensive in CPU time, but

always results in the best overall solution. Hence, this is appropriate if there are only two or three candidates.

Depending on the selection heuristics, however, it is still possible that these top-down selection methods result in a final circuit schematic which cannot achieve the specified performance. To avoid this iteration over topologies, one can use techniques from interval analysis to determine the performance range of a higher-level block from the performance ranges of the subblocks [LEE_90]. In this way, it can be guaranteed that, given the top-level specifications, all selected circuit schematics can meet these specifications.

- *bottom-up topology construction* :
these methods use general principles to construct an appropriate topology from scratch for each application, by assembling elementary blocks such as transistors [DEK_84, RES_84, ROY_80]. However, up till now, these methods have not yet proven to be practically useful, as they are usually struck by the large number of constraints generated for each design.

- *non-fixed-topology systems* :
these systems facilitate the fast, nearly automatic inclusion of new topologies by the designer himself. Of course, selecting from fixed, predefined alternatives [DEG_87c, HAR_89b, KOH_90, etc.] is the most straightforward topology generation method, since all required knowledge has to be derived only once and is then stored together with the topology. This can be appropriate for many applications. However, the derivation of the topology-specific knowledge is a long, tedious and error-prone job, which has to be simplified by CAD tools. In addition, many analog high-performance applications require minor or major modifications to existing schematics or even totally new schematics in order to meet the stringent performance requirements. Therefore, analog design systems must not be limited to a fixed set of topologies, but additionally need to have some facilities to easily introduce new schematics and the corresponding knowledge into the system, by the designer himself or at least by the tool engineer [GIE_90a].

Remember also that a topology is defined as an interconnection of blocks from a lower level in the hierarchy. Depending on the number of hierarchical levels used, one can distinguish :
- **flat-schematic approaches**, such as IDAC [DEG_87c], which express a block (e.g. an operational amplifier) directly in terms of transistors. Selection of a topology then consists of the selection of a transistor schematic.
- **hierarchical approaches**, such as OASYS [HAR_89b], which express a block in terms of subblocks, and these subblocks in terms of subsubblocks, and so forth until the transistor level is reached. For example, an operational amplifier consists of current mirrors, a differential pair, etc., which are themselves composed of transistors. For larger analog modules, more hierarchical levels can be

distinguished. These methods require the selection of an appropriate topology (also called design style) at every hierarchical level.

An hierarchical approach has several advantages. First of all, at the higher levels one is not overwhelmed by transistor-level details. Instead, a block is represented in a more abstract way as an interconnection of abstract subblocks. The design details appear only gradually while moving down in the hierarchy. In addition, since an appropriate topology has to be selected at each level in the hierarchy, many transistor schematics correspond to the same high-level topology [HAR_87]. Hence the designer has many more possibilities in choosing an appropriate schematic for the given specifications than with flat-schematic approaches. Flat-schematic approaches obviously are more limited in the number of possible schematics. To still cover a wide range of performance specifications, they need a large database of schematics and/or the flat schematics have to be as generally and widely applicable as possible. The first approach is used in IDAC which has presently more than 50 schematics [DEG_89]. The second approach is used in OPASYN with a small database of generic, widely applicable opamp schematics [KOH_90].

2.2.3. Classification based on the sizing method

The second task is the sizing of the selected circuit topology for the given specifications and technology data. Depending on the **technique used for the sizing or the dimensioning,** one can distinguish :

- *optimization-based systems :*
  they consider the sizing as a constrained optimization problem. Depending on how they evaluate the circuit performance in the inner loop of the optimization, they can be subdivided in :
  - * *simulation-based methods :*
    they perform a numerical simulation of the circuit at each optimization step.
  - * *equation-based methods :*
    they model the circuit behavior with analytic equations which are evaluated at each optimization step.
- *knowledge-based systems :*
  they perform the sizing by explicitly representing and exploiting knowledge. Depending on the kind of knowledge used, they can be subdivided in :
  - * *methods based on general principles :*
    they attempt to exploit universal methods that underly all circuit design activities, such as Kirchhoff's laws, to design circuits from scratch.
  - * *methods based on domain-specific knowledge :*
    they exploit detailed analog design knowledge (topological and analytical knowledge, rules of thumb, heuristics, simplified models, etc.) to perform the circuit sizing. This knowledge can be topology-specific or more generally applicable to various topologies. At the same time, the internal representation

of this knowledge can widely vary (e.g. rules, design plans, hard-coded procedures...).

Methods based on general principles [DEK_84, RES_84, ROY_80] appear to be too complicated for practical real-life designs. More pragmatic are the knowledge-based approaches [DEG_87c, HAR_89b, etc.] that utilize the existing analog design knowledge to resolve the degrees of freedom in the design and complete the sizing. However, these methods are usually biased towards some implicit optimization goal (such as minimum power) and are less suited for other goals (such as minimum noise). Also, due to the many heuristic decisions, they only produce acceptable initial solutions, which still have to be fine-tuned afterwards (by means of local optimization). Optimization-based methods [GIE_90b, KOH_90, NYE_88, etc.], on the other hand, have explicit optimization goals and inherently manage all trade-offs in the design. The solution of local optimization techniques of course depends on the selection of the initial starting point. Global optimization techniques, on the other hand, can yield a solution close to the global optimum, but at the expense of larger CPU times. These CPU times, however, can be reduced at the expense of some loss of accuracy by using analytic circuit models [GIE_90b, KOH_90] instead of doing a full simulation at each iteration [NYE_88].

Note also that the above distinction between optimization-based and knowledge-based techniques is not strict and that many design systems use some combination of both techniques. Indeed, simplified analytic models used during optimization are part of the analog design knowledge in addition to the more heuristic, intuitive knowledge. Also, knowledge-based systems often use some form of optimization to determine certain variables, which are difficult to fix by heuristics or which simply have to be obtained by iteration.

2.2.4. Classification based on the layout generation method

The third task is to lay out the sized circuit for the given technology data. Depending on the **technique used for generating the layout,** one can distinguish :
- *parameterized-layout systems :*
  they have stored the geometry information for a circuit in parameterized form or even in fixed form (e.g. when using analog standard cells).
- *custom-layout systems :*
  they generate the layout from scratch for every application.

It is quite obvious that after an application-tailored sizing, the circuit also has to be laid out in a full-custom-like style in order to obtain the best overall performance at the lowest cost (in power and area) [COH_90, GAR_88, KAY_88a, RIJ_89, etc.]. In addition, for many high-performance applications, the layout is critical to ever meet the stringent specifications, thereby excluding all semi-custom layout styles.

All the automated analog design programs presented in the literature can be classified in one or more of the above categories. In the next section, an example analog design system is discussed in more detail. This design system, which customizes the circuits to each application, uses symbolic simulation techniques to simplify the introduction of new schematics and thus to create an open analog design environment.

2.3. An automated design methodology for analog modules

2.3.1. Description of the ASAIC methodology

To illustrate the basic concepts in analog module generation and to motivate the development of symbolic analysis techniques, we now discuss one particular methodology for the automated design of analog functional modules in greater detail. This methodology, being implemented in the ASAIC (Automatic Synthesis of Analog Integrated Circuits) system, has been developed in the ESAT-MICAS group of the Katholieke Universiteit Leuven, Belgium [GIE_86, GIE_90a, GIE_90c, GIE_90d, SAN_88a, SWI_89, SWI_90, WAL_91]. It **combines symbolic simulation, numerical optimization and knowledge-based techniques to allow optimal circuit sizing and fast inclusion of new topologies** [GIE_90a]. The design methodology is schematically shown in Fig. 2.2. It covers the whole analog module design path from topology generation over analytic modeling and circuit sizing down to layout. The sizing of the modules from performance and technology specifications is then generally formulated as a constrained optimization problem, based on an analytic design model of each module. This model contains analytic equations describing the behavior of the analog module, as well as general and circuit-specific knowledge, and is automatically generated as far as the AC characteristics is concerned. The methodology will be illustrated for the design of operational amplifiers in section 2.4 below, but is generally applicable to any analog module if the corresponding analytic models are available. According to Fig. 2.2, the design can then be carried out fully automatically, relying on the built-in knowledge and analytic models. However, the system additionally provides features for the designer to guide the design session interactively or to introduce new schematics and design knowledge himself. The methodology is now described below.

At the top, the user (or a high-level synthesis program) enters the module design system with specifying the technology process and performance specifications for some type of analog functional module. The **topology generation** program HECTOR then decides on an appropriate topology to implement the module. This is implemented as an expert system that inspects its knowledge base and proposes an appropriate topology for the given specifications and technology [DON_90, SWI_89]. Note that these topologies are defined hierarchically as an interconnection of topologies on a lower level in the hierarchy, and that a topology has to be selected at each level in the hierarchy down to the transistor level. The user can accept this topology, or he can enter a new one (or some part of it) himself. This last capability is one of the strengths of the design system.

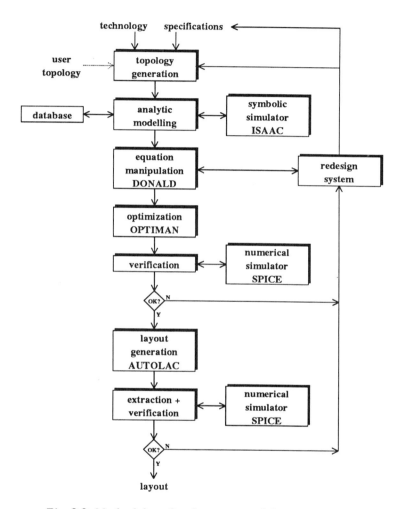

Fig. 2.2. Methodology for the automated design of analog functional modules.

The result is then passed to the **analytic modeling** routine which generates the analytic design model for the topology. These analytic models have to be independent of the context the topology is used in. They are also built up hierarchically resembling the topological hierarchy [CRU_90]. If no new topology has been entered, the analytic model is retrieved from the design system's database.

If a new schematic has been entered, a building block recognizer tries to recognize known (sub)circuits to ease the analytic characterization [SWI_89]. However, if the new topology (or some part of it) is not present in the database, a new analytic model is generated. This is accomplished in an automated way for the AC characteristics by use of the symbolic simulator ISAAC (Interactive Symbolic Analysis of Analog Circuits) [GIE_89c, SAN_89]. For the time-domain and the large-signal characteristics, the designer still has to provide the design equations himself. The AC characteristics, however, typically cover 75% of all characteristics for an analog module such as an operational amplifier. The automation of the derivation of the AC characteristics thus results in a substantial reduction of the time needed to introduce a new schematic into the system. In addition, the designer can always add some knowledge about the new circuit in the form of additional analytic relations. The new model is then stored into the database such that it can be reused in future design problems.

Next, the **design equation manipulation** program DONALD (Design equatiOn maNipulation for Analog circuit Design) [SWI_90] transforms the analytic circuit model, which is internally represented as an equation network, into a solution plan which is appropriate for the circuit optimization. In addition, DONALD has built-in features to deal with variations in technological and environmental parameters. The **optimization** program OPTIMAN (OPTIMization of Analog circuits) [GIE_89a, GIE_90b] then sizes all circuit elements based on this solution plan, to satisfy the performance specifications for the given technology data. The degrees of freedom in the design are used to minimize a user-supplied cost function, e.g. a large output swing, low quiescent power, a fast settling time and low distortion for an output buffer.

The design is then verified in the **verification** routine. The use of analytic circuit models and design rules speeds up the optimization and produces close-to-optimal results. However, because of any simplifications and rules used, a final verification has to be carried out. This verification consists of several tests, such as controlling the operating region of each transistor, and a final performance check by means of a numerical simulator such as SPICE [VLA_80b] or SWAP [SWA_83]. If the design is accepted, the circuit schematic and the device sizes are passed to the layout program. If some specifications are violated, the **redesign system** is entered. Based on the failure information, DONALD then traces the characteristics that are responsible for the failure. Consequently, a backtrace is carried out to tighten some internal specifications, to modify or reselect a new topology or to trade off some input specifications.

The **layout generator** AUTOLAC (AUTOmatic Layout of Analog Circuits) [SCH_88] then lays out the analog module, starting from the circuit schematic and size information and taking into account requirements such as minimum area, capacitive coupling avoidance, matching, balancing, etc. Parasitic elements are then extracted from the layout and a final verification is performed. If the design is not acceptable after layout, the circuit is resized or eventually a new topology is required. Finally, if the design is accepted, a complete behavioral model is generated to simulate the module at the system level.

2.3.2. Advantages and limitations of the ASAIC methodology

The **main features of the ASAIC design methodology** can now be summarized as :
- it is quite general and independent of any particular topology or technology. All topology-specific information is combined into the analytic design models, and all technology-specific information is contained in the appropriate technology files.
- it can design analog modules fully automatically, but it also allows the designer to include new schematics and design knowledge himself. In this way, ASAIC targets both analog system designers and analog circuit designers, in an attempt to increase the acceptability of CAD and DA tools by the analog design community.
- the inclusion of new topologies is largely facilitated due to the link with the symbolic simulator ISAAC [GIE_89c]. This symbolic simulator automatically generates all AC characteristics for the new topology. These features make ASAIC to a flexible open design system, that is not limited to a fixed set of circuit topologies [GIE_90a]. This sldo motivates the detailed discussion of symbolic simulation techniques in the following chapters.
- it optimizes each module according to the specifications of the application. The sizing in OPTIMAN is based on a global optimization method [GIE_90b], which implicitly takes care of all degrees of freedom and trade-offs in the design and obtains a solution close to the global optimum (of course at the expense of larger CPU times). The system, however, also works with any faster local optimization method (with an initial starting point provided by DONALD). The layout is also generated in a customized way.
- the strength of the methodology lies in the automatic generation of the AC equations, and the automatic manipulation of all analytic design equations. This emulates the manual design procedure of analog designers: they initially design a circuit based on (manually derived) first-order formulas and then fine-tune the result with a numerical simulator such as SPICE [VLA_80b]. Our methodology is similar, but most of the design equations are derived automatically in a fast and correct way and can thus be much more accurate. Moreover, a designer cannot master a multitude of performance characteristics and their trade-offs at the same time, whereas a general optimization program can [GIE_90a].

The proposed methodology, as all approaches, also has **some limitations** :
- first of all, it is only applicable to designs where analytic design equations can be generated. The present symbolic simulators can generate small-signal frequency-domain characteristics for both linear and weakly nonlinear circuits only, not yet

time-domain or large-signal characteristics (which is much more difficult to automate). Hence, these latter expressions still have to be generated by the designer. However, for all other analog design systems presented up till now, all characteristics of the circuit (including the AC characteristics) have to be derived by the designer (or the tool creator) himself. **The ASAIC design system, on the other hand, automates large part of this analytic model acquisition process, as the AC characteristics, which typically cover 75% of all characteristics for an analog block such as an operational amplifier (see appendix A), are derived automatically and to any desired accuracy.** The automation of this 75% is thus already a big leap forward towards a truly non-fixed-topology system, as it greatly reduces the time needed to introduce a new circuit into the design system. For other analog modules, such as comparators or oscillators, the design methodology is applicable as well, but probably larger part of the analytic design model has to be provided by the designer himself. In addition, note that the methodology is quite general in that the analytic design model may contain anything from simplified analytic equations to more complicated subroutines (e.g. a partial numerical simulation) and thus that a complete closed analytic description of the module is not necessary.

• secondly, the analytic equations are always approximations of the real circuit behavior. The resulting design has to be verified carefully, and possibly has to be redesigned. However, since the design equations can be of virtually any accuracy [GIE_89c], the number of design iterations is reduced to one or two as opposed to the multitude of iterations during manual design. By using approximated formulas, the optimization is definitely faster than if a full simulation were performed at every iteration [GIE_90b].

2.4. The methodology applied to a simple example

To clarify the overall design flow and the main routines within the ASAIC system, the methodology is now illustrated by means of a practical design example. For the sake of clarity, however, the example has been highly simplified. It is only intended to exemplify the main ideas in the ASAIC approach.

The most frequently used analog building block is the operational amplifier. It forms the cornerstone of numerous analog modules, such as A/D and D/A converters and filters. An efficient design of operational amplifiers is thus crucial for many analog applications. In addition, operational amplifiers are well understood and characterized nowadays [ALL_87, GRA_84, LAK_91, SAN_88b]. This explains why operational amplifiers are the first target circuit class to test many analog design systems. For this, we also select the design of an operational amplifier as example.

The task is then to design a multi-purpose operational amplifier driving on-chip loads. The technology is given to be the MIETEC 3-μm n-well CMOS process. The specifications imposed on the opamp are :
- a maximum load of 100 kΩ and 10 pF
- a minimum gain-bandwidth of 1 MHz
- a minimum gain of 60 dB
- a minimum phase margin of 60°
- an output range of ± 2 V (for ± 2.5 V power supply voltages).

The design objective is to minimize the power and area consumption. The ASAIC system will then proceed by first selecting a suitable topology, then analytically characterizing this topology, manipulating the resulting design equations, optimally sizing the circuit, verifying the resulting design and finally laying it out and verifying it again. These main steps are now described separately in somewhat more detail.

2.4.1. Topology generation

The first task is the generation of a candidate topology, which most likely can meet the performance specifications. In the ASAIC methodology, this topology is generated by the top-down topology selection program HECTOR, which has currently been prototyped [DON_90, SWI_89]. It is based on a library of frequently used topologies from several hierarchical levels. Depending on the input specifications and the technology data, an expert system ranks the possible topologies according to its built-in rules and suggests the most viable one. This is then repeated at each hierarchical level, resulting in a local optimization at each level. In the case of an operational amplifier for instance, first a decision is taken whether to use a one-stage or a two-stage buffered/nonbuffered opamp topology. Next, possible generic circuit topologies for a stage or a buffer are presented consisting of building blocks such as differential pairs, current mirrors, etc. This process continues until finally a topology at the transistor level is assembled. The program can run automatically, but it can as well be used interactively by a designer. In the latter case, the expert system is to be considered as an intelligent advising design assistant, which leaves the final decision to the user. The user can always refuse a proposed topology and enter a new one himself [SWI_89].

For the above example of the multi-purpose CMOS opamp, the combination of the input specifications (e.g. relatively high gain and moderate gain-bandwidth) asks for a nonbuffered two-stage amplifier at the opamp level. Next, a decision is taken for the subblocks: both the differential pair, the current mirror, the transconductance amplifier, the current sources and the compensation chain can be implemented in simple form for the above specifications. This selection process is conceptually illustrated in Fig. 2.3. It results into the final schematic shown in Fig. 2.4.

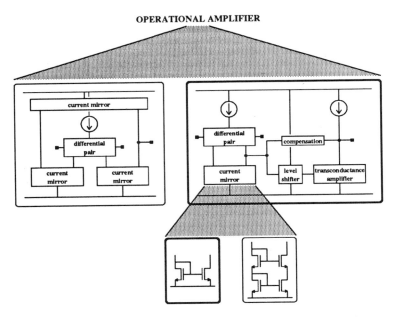

Fig. 2.3. Hierarchical composition of a two-stage opamp topology.

2.4.2. Analytic modeling

In the ASAIC methodology, all modules are characterized by an analytic design model [GIE_86, GIE_89a, GIE_90b]. This analytic model in general contains a description of the circuit topology, design equations relating the circuit performance to the design variables, additional analog expert knowledge about the circuit and other general or circuit-specific design constraints. As opposed to other approaches such as IDAC [DEG_87c] and OPASYN [KOH_90], however, the generation of the analytic model for a new topology is highly automated by using the symbolic simulator ISAAC [GIE_89c], which automatically generates all AC characteristics for the new circuit, thus largely reducing the effort to include a new topology into the system.

In general, an analytic model consists of three different classes of characteristics:
- *AC characteristics* are generated automatically by the symbolic simulator ISAAC [GIE_89c], which analyzes both time-continuous and time-discrete (switched-capacitor) lumped, linear or linearized (small-signal) circuits in the frequency domain and returns analytic expressions with the complex frequency and (all or part of) the circuit elements represented by symbols. Since for semiconductor circuits the exact expressions are usually lengthy and complicated, a heuristic approximation algorithm has been built into the program, which simplifies the

expressions based on the relative magnitudes of the elements and a user-supplied error percentage and which returns the dominant terms in the result only. The ISAAC program and symbolic analysis techniques in general are thoroughly discussed in the next three chapters. Note, however, that - besides for automatically generating the AC characteristics for the analytic model of a circuit - a symbolic analysis program is also valuable as an interactive stand-alone tool for instruction and designer assistence, since it provides the user with analytic insight into a circuit's behavior.

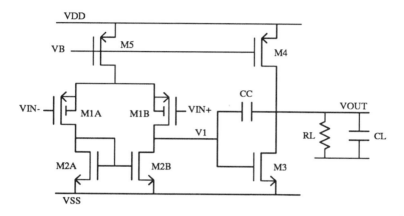

Fig. 2.4. Schematic of the CMOS Miller-compensated two-stage operational amplifier.

Example
For the CMOS two-stage Miller-compensated opamp schematic of Fig. 2.4 selected for our example, the following AC expressions describing the gain-bandwidth and the phase margin have been obtained by means of ISAAC :

$$GBW = \frac{g_{m1}}{2\pi\ C_C} \tag{2.1}$$

$$P_2 = \frac{g_{m3}}{2\pi\ [C_L\ (1\ +\ C_1/C_C)\ +\ C_1]} \tag{2.2}$$

$$z = \frac{g_{m3}}{2\pi\ C_C} \tag{2.3}$$

$$PM \approx 90° - atan(\frac{GBW}{p_2}) - atan(\frac{GBW}{z}) \tag{2.4}$$

where C_1 is the internal capacitance on the output node (V_1) of the first stage :

$$C_1 = C_{GS3} + C_{DB1} + C_{DB2} \tag{2.5}$$

Additional equations relating the small-signal parameters in these formulas to sizes and currents are added by DONALD. They depend on the transistor model used. For example, the relationship between the transconductance, the current and the saturation voltage (or the current and the aspect ratio) of a MOS transistor in saturation and strong inversion is given by :

$$g_{mi} = \frac{2 I_i}{(V_{GS}-V_T)_i} = sqrt(2 \beta_i I_i) \tag{2.6}$$

For a complete analytic description of all characteristics for this two-stage operational amplifier, the reader is referred to appendix A.   ◆

- The *time-domain or large-signal characteristics*, such as the settling time or the slew rate, still have to be provided by the designer himself (as in all other existing analog design tools). The present symbolic simulators are not yet able to automatically derive these expressions (if it will ever be possible).
- Finally, the designer can also include *heuristic design rules* into the analytic model. For example, in order to reduce the systematic offset voltage in the opamp of Fig. 2.4, the designer can impose the following condition :

$$(V_{GS}-V_T)_2 = (V_{GS}-V_T)_3 \tag{2.7}$$

All these expressions are combined into the analytic design model of the analog module. Note that these expressions are usually analysis equations, expressing the performance characteristics in terms of the design parameters (e.g. bias currents, device sizes). For the sizing, these analysis equations have to be inverted into design equations, which inversely express the design parameters in terms of the given performance specifications. This inversion, however, is usually underconstrained and the remaining degrees of freedom in the design have to be resolved by the design system in one way or another. In the ASAIC methodology, this is accomplished by formulating the sizing as a constrained optimization problem based on the circuit's analytic model. For this, however, the analytic model first has to be converted into the appropriate form for the optimization. This is performed by the equation manipulation program DONALD [SWI_90].

2.4.3. Design equation manipulation

Once the analytic circuit model has been generated, the equation manipulator DONALD [SWI_90] converts this model into a solution plan for the optimization by extracting the independent design variables and constructing a computational path out of the model. The analytic circuit model consists of a large set of (linear and nonlinear, explicit and implicit) equations, all expressing some relation between the design variables. As design variables, all node voltages, currents and element values in the circuit are taken, as well as all performance characteristics. Most of these variables, however, are interrelated because of general constraints (such as Kirchhoff's laws), circuit-specific constraints (such as analysis equations relating the circuit performance to the circuit elements, or matching information) and designer constraints (such as offset-reduction rules). Hence, in order to allow an efficient optimization, a minimal set of **independent variables** has to be determined. This task is performed by the equation manipulator DONALD [SWI_90]. **The number of independent variables is equal to the degrees of freedom in the design**, i.e. the difference between the total number of variables in the design and the number of equations. The number of degrees of freedom is usually quite large, since there are many more variables than equations in practical design examples. **The independent variables can now be interpreted as the fundamental design parameters, on which all characteristics of the circuit depend by means of the given analytic equations. If values are assigned to all independent variables, then all other variables and hence the total circuit behavior can be calculated.** Obviously, the independent variables are the parameters to tune and optimize the circuit performance. The optimal values for the independent variables will then be selected by the optimization program OPTIMAN [GIE_90b]. Note that the set of independent variables is not unique, but some sets are computationally more advantageous than others. Note also that any variable, even a specification, can be used as independent variable, not only transistor widths, lengths and currents.

DONALD internally represents the analytic circuit model as a bipartite graph, in which the nodes either represent equations or variables [SWI_90]. The program then automatically (or interactively) constructs a possible set of independent variables out of this bipartite graph. The program also finds out how all the other, dependent variables are to be computed out of the independent ones. This results in a computational path, which essentially is a sequential list of calls to numerical routines to calculate all the dependent variables. Often, some dependent variables are so strongly interrelated that they can be calculated only by solving a set of equations. However, the numerical routines are strong enough to handle arbitrary systems of nonlinear equations.

Example

These ideas are now illustrated with a simple example which is part of the design example of the CMOS multi-purpose opamp. Fig. 2.5 shows the bipartite graph representing the following two (highly simplified) equations, which allow to design the input transistors :

$$GBW = \frac{g_{m1}}{2\pi \, C_C} \tag{2.8}$$

$$g_{m1} = \frac{2 \, I_1}{(V_{GS}-V_T)_1} \tag{2.9}$$

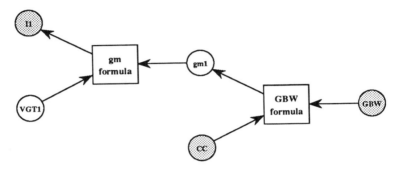

Fig. 2.5. Illustration of bipartite graph representation and computational path construction.

In this simplified example, there are 5 variables $\{GBW, g_{m1}, C_C, I_1, (V_{GS}-V_T)_1\}$ and 2 equations $\{GBW\_formula, g_m\_formula\}$. Hence, there are 3 degrees of freedom and any set of 3 out of the 5 variables can be selected as the set of independent variables. If for example I_1, GBW and C_C are chosen as independent variables, the following computational path is constructed to compute the dependent variables g_{m1} and $(V_{GS}-V_T)_1$ out of the independent ones :

$$g_{m1} \text{ out of } \{GBW, C_C\} \text{ by means of } \{GBW\_formula\} \tag{2.10}$$

$$(V_{GS}-V_T)_1 \text{ out of } \{I_1, g_{m1}\} \text{ by means of } \{g_m\_formula\} \tag{2.11}$$

Note that in this example the gain-bandwidth is considered as independent variable, and the corresponding value of $(V_{GS}-V_T)_1$ is calculated. One could as well select $(V_{GS}-V_T)_1$ as independent variable, and calculate the resulting GBW value. The use of GBW as independent variable has the advantage that the minimum gain-bandwidth constraint can be satisfied in a natural way, whereas the use of $(V_{GS}-V_T)_1$ as independent variable allows to control the region of operation of the transistor in a natural way. For DONALD, both possibilities are equally valid, whereas a designer prefers currents and aspect ratios as independent variables.   ◆

Now a set of independent design variables for the CMOS Miller-compensated two-stage opamp of Fig. 2.4 is selected. We need a variable for each unknown external bias current and voltage in the circuit and for the value of each unknown element (e.g. C_C) in the circuit. In addition, each transistor in the circuit has to be characterized by 2 variables out of $\{I, V_{GS}-V_T, W/L\}$ and 1 variable out of $\{W, L\}$. (In our example, however, the transistor lengths are fixed.) This leaves us with 15 variables for the opamp of Fig. 2.4. This large number of variables is then reduced to an independent set by applying Kirchhoff's current law as well as matching and designer constraints to the circuit. This then results in the following set of independent variables $\{I_1, I_3, (V_{GS}-V_T)_1, (V_{GS}-V_T)_2, (V_{GS}-V_T)_4, C_C\}$. The resulting computational path is omitted here. More details on the DONALD program can be found in [SWI_90].

2.4.4. Design optimization

The set of independent variables and the computational path are then passed to the optimization program OPTIMAN, which sizes all elements to satisfy the performance constraints, thereby optimizing a user defined design objective [GIE_90b]. During this optimization, the independent variables are varied. The performance specifications are treated either as objectives to be minimized (or maximized) or as (equality or inequality) constraints. For opamp designs, the goal function could be a weighted combination of characteristics such as power, chip area, noise...

The OPTIMAN program is described in more detail in chapter 6. The optimization algorithm used is simulated annealing [KIR_83]. This is a general and robust global optimization method, based on random move generation and statistical move acceptance. Since it also allows up-hill moves, a solution close to the global optimum can be found starting from any initial solution at the expense of a large number of function evaluations. However, the CPU times are kept acceptable by the application of an efficient and adaptive annealing schedule [CAT_88].

The general optimization loop within OPTIMAN is then as follows [GIE_90b]: the program statistically selects new values for the independent variables in a move range around the present value, it calculates the dependent variables by means of the computational path, checks if the corresponding design satisfies all boundary conditions and constraints, calculates the goal function and statistically accepts or rejects the new state. This loop is executed until convergence occurs at lower temperatures, while the move range is gradually decreased with decreasing temperature.

Example

This is now illustrated for the design example of the CMOS multi-purpose opamp of Fig. 2.4. Table 2.1 compares the optimized values for the set of independent variables $\{I_1, I_3, (V_{GS}-V_T)_1, (V_{GS}-V_T)_2, (V_{GS}-V_T)_4, C_C\}$ to the values obtained by an expert designer for the same set of specifications [SAN_88b, GIE_90d]. Both results are in quite good agreement, but OPTIMAN allows to obtain a close-to-optimal solution also for more complicated design problems.   ♦

	OPTIMAN	expert designer
I_1	1.0 μA	0.63 μA
I_3	22 μA	25 μA
$(V_{GS}-V_T)_1$	0.21 V	0.2 V
$(V_{GS}-V_T)_2$	0.28 V	0.2 V
$(V_{GS}-V_T)_4$	0.35 V	0.5 V
C_C	1.3 pF	1 pF

Table 2.1. Comparison of values for the set of independent variables obtained by OPTIMAN and by an expert designer for the design of a CMOS two-stage Miller-compensated opamp.

2.4.5. Verification

The sized schematic is then functionally verified. This verification consists of :
- several tests to check the resulting design. For example, the operation region of all transistors is checked, for MOS transistors it is investigated whether short-channel effects occur or not, etc.
- several numerical simulations. The use of simplified analytic models highly speeds up the optimization. However, to fully verify the resulting design, all second-order effects have to be included and all components have to modeled as accurately as possible. To this end, a numerical simulator (such as SPICE [VLA_80b] for time-continuous circuits or SWAP [SWA_83] or SWITCAP [FAN_83a, FAN_83b] for switched-capacitor circuits) has to be used. However, to fully verify say an operational amplifier, the verification program has to run several numerical simulations: transient analysis for the settling time and the slew rate; DC transfer characteristic; AC analysis for the open-loop gain and the stability, the PSRR, the CMRR... ; noise analysis; distortion analysis; etc. Moreover, these analyses have to be carried out at nominal device characteristics and with technology, temperature and bias variations. The verification task consists clearly of more than one simulation. Actually, it results in a complete datasheet for the designed module.

Depending on the verification results, the design is then accepted or rejected. In the latter case, the redesign system is entered which traces for the failing characteristics and then takes an appropriate corrective action, such as changing the internal specifications or modifying the topology.

2.4.6. Layout generation

If the design is accepted by the verification program, the circuit has to be laid out in a customized way. In the AUTOLAC program [SCH_88], which is currently still under development, this is accomplished in four steps : the circuit is first divided in structural entities, which are then optimally placed and interconnected, after which the layout is compacted. In this way, a near-optimal layout can be generated for every application, taking typical analog constraints into account.

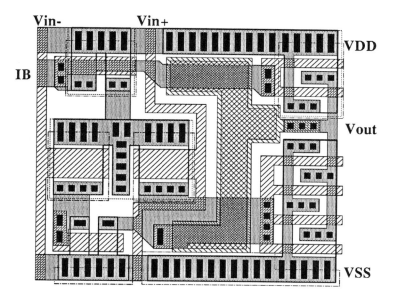

Fig. 2.6. Layout of the multi-purpose CMOS two-stage Miller-compensated operational amplifier.

The structural entities are functional groups of elements, such as two or more matching transistors, which are taken together in order not to overconstrain the placement routine. These entities can be laid out in several ways, but the actual layout form is only decided during the placement. During this placement, the overall area and interconnection is minimized, under typically analog boundary conditions such as symmetry requirements. The router also has to incorporate typically analog features, such as crosstalk avoidance, parasitic capacitance reduction on critical nodes, separation of sensitive and noisy nodes, etc. Finally, the analog compactor has to maintain all analog constraints in the layout.

The layout for the CMOS Miller-compensated two-stage opamp which is the final realization of the CMOS multi-purpose opamp is shown in Fig. 2.6. After the generation of the layout, a layout extractor extracts the circuit schematic from the layout with all layout parasitics included. This extracted circuit is then again verified. If it satisfies all performance requirements, the design task is completed and the layout is returned to the user (or the high-level synthesis program). At the same time, a behavioral or macromodel is generated for simulating the designed module at the system level. If the extracted circuit fails to meet some specifications, the redesign system is entered again and the design has to be modified in an iterative loop.

If the overall design is accepted, the chip is processed. The resulting parameter values and performance characteristics of the opamp designed in this example, together with measurement results from integrated samples, are summarized in appendix A.

Conclusion

In this section, the ASAIC methodology for the automated design of analog functional modules [GIE_90a, SWI_89, GIE_90d] has been illustrated by means of a simple example, and the different subtasks in the methodology have been exemplified. With respect to the present implementation state of the ASAIC system, the programs ISAAC [GIE_89c], OPTIMAN [GIE_90b] and DONALD [SWI_90] have already been completed, the topology selector HECTOR has been prototyped [DON_90] while AUTOLAC [SCH_88] still has to be completed with an analog router.

2.5. Discussion of and comparison with other analog design systems

In the past five years, many other analog circuit design systems and programs have been published as well. The most viable and matured systems (as judged by publication in IEEE journals) will now be discussed in some more detail in subsection 2.5.1. This overview, however, by no means intends to be complete. It only provides a representative sample of existing systems which have evolved far enough to produce practical circuits. These different methods and the ASAIC system, which has been described and illustrated in the previous two sections, are then compared in detail in a summarizing table in subsection 2.5.2. In this way, the reader is provided with a realistic view of the present state of the art in the field of automated analog module or circuit design.

2.5.1. Discussion of other analog design systems

2.5.1.1. ADAM

The ADAM circuit design environment, developed at CSEM, Neuchâtel, Switzerland, consists of the IDAC and ILAC programs [DEG_84, DEG_87c, DEG_89, DEG_90, RIJ_89]. ADAM performs sizing and generates a custom layout for several fixed schematics contained in the system's library. It is now commercialized as the AutoLinear package by SCS/Mentor Graphics [WEB_89] and is the first really commercial analog design automation tool. The principle block diagram of the ADAM environment is shown in Fig. 2.7.

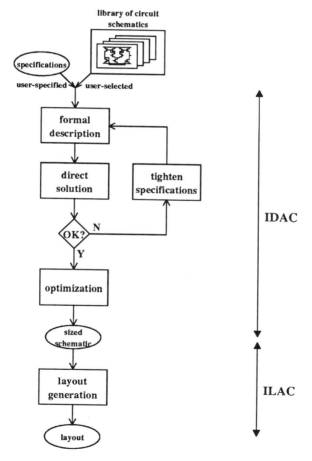

Fig. 2.7. Block diagram of the ADAM analog circuit design environment.

The IDAC program [DEG_84, DEG_87c, DEG_89] contains a library of more than 50 flat-level circuit schematics, such as operational amplifiers, voltage and current references, oscillators... It takes as input building block specifications and technology data. The user also has to specify some values for the degrees of freedom in the design and he has to select one or more circuit schematics himself. IDAC then first predistorts the user specifications for each selected schematic by means of some circuit-specific first-order expressions in order to account for technology, temperature and bias variations. Based on the modified specification values, IDAC then performs a first-cut sizing of the selected schematic(s) by executing the corresponding formal description(s). This description is based on the analytic circuit equations which describe the performance of each schematic. These equations have been derived after an in-depth analytic study and have been inverted explicitly by an experienced designer, to yield a straightforward step-by-step design procedure to calculate the device dimensions and bias currents from the specification values for each particular schematic. The inversion from analysis equations into design equations is carried out by means of heuristic design knowledge, mainly directed towards low power and low area, but any remaining degrees of freedom still have to be determined by the user.

The IDAC program then verifies each sized schematic with a built-in analyzer dedicated to each class of building blocks. If the design fails, the failing specifications are tightened and the design is restarted. If it still fails, a failure message is returned to the user. If the design satisfies, the program returns the sized schematic together with a complete data sheet. Typical CPU times for this sizing are a few seconds on a VAX 780. Starting from this initial knowledge-based design solution, IDAC can also perform a steepest-descent-like optimization in order to further improve the total circuit performance.

The ILAC program [RIJ_89] then completes the design by automatically generating the layout in a macrocell place-and-route style, while handling typically analog layout constraints such as symmetry, matching and clustering requirements. The designer can also modify the layout interactively at the symbolic level.

IDAC has a large library of different classes of circuits and can size practical circuits very rapidly, at least on the conditions that the circuit you want is in the library and that your specifications are within the IDAC target scope (low power, moderate frequency range). These conditions, however, are not always fulfilled, and for agressive specifications (such as high bandwidths) IDAC often cannot obtain any solution. The program also supports only very little form of hierarchy. In addition, it does not explicitly have any underlying framework that exists independently of the circuit knowledge. Instead, it just consists of a set of programs, one for each schematic. The failure recovery is also limited to tightening the failing specifications without looking for the real failure causes or without automatically changing the topology. It is left to the designer to change some of the values entered for the degrees of freedom in the design or to select another schematic. Finally, it requires an enormous effort to introduce a new schematic into the system, since all analytical and heuristic knowledge about the schematic has to be collected and hard-coded into the circuit's formal description by an expert designer, who has to be familiar with IDAC's internal organisation. This is a time-consuming job to be repeated for every circuit in the program's database.

In this context, it is interesting to mention that the ADAM system has been announced recently to evolve in the near future to an open framework in which experienced designers can add their own knowledge and new circuits by means of a tableau description of the circuit, indicating the weight of the device contributions to several performances [DEG_89, DEG_90]. In the future, this tableau is expected to be generated automatically by the symbolic simulator SYNAP [SED_88].

2.5.1.2. ACACIA

The ACACIA analog design environment, developed at Carnegie Mellon University, Pittsburg, U.S.A., consists of the synthesis framework OASYS, the layout generator KOAN/ANAGRAM and a graphical interface [CAR_89, COH_90, GAR_88, HAR_89b]. ACACIA hierarchically generates and sizes a topology based on the input specifications and then produces a custom layout. The principle block diagram of the environment is shown in Fig. 2.8.

OASYS [HAR_89b] is aimed to be a framework for analog-behavior-to-sized-schematic synthesis for some specified classes of analog building blocks. It supports a knowledge-based synthesis approach, based on the observation that analog design experts usually dispose of mature plans of how to attack the design of common circuits. OASYS then relies on three main ideas: hierarchical decomposition, topology (or design style) selection and specification translation. All circuit topologies are built up hierarchically as an interconnection of lower-level blocks. Transistors are only found at the lowest level. At the program input, the user enters the specifications and the technology data. The synthesis is then accomplished by alternately selecting a design style (a topology) for the current block from a fixed set of alternatives, for example a simple or a cascoded current mirror, and translating the specifications of the present block into specifications for the subblocks of the selected design style. This process is repeated down the hierarchy until the transistor level is reached. The specification translation however is a nonunique process, which can only be realized by exploiting design knowledge and heuristics to eliminate all degrees of freedom (sometimes with some simple iteration). In OASYS, this knowledge is implemented in the form of a simple design plan, attached to each topology. The translation is then performed by executing the design plan for the selected topology and for the given specifications. If the design plan fails somewhere, a plan fixer is entered which tries to modify the design plan or tracks back to modify the specifications or changes the topology up the hierarchy. The resulting design is then verified by numerical simulation.

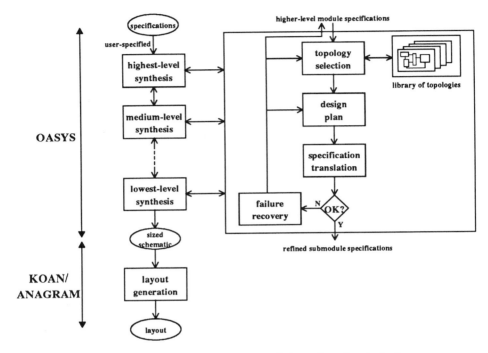

Fig. 2.8. Block diagram of the ACACIA analog circuit design environment.

The KOAN/ANAGRAM layout program [COH_90, GAR_88] then generates a custom layout for the OASYS-sized schematic. It consists of an analog-oriented component placement and routing, which considers symmetry conditions, crosstalk avoidance, device merging, etc.

The present version of OASYS [HAR_89b] contains two opamp and one comparator design style at the top level. The program can synthesize circuits very rapidly due to the quite straightforward design plans. Typical CPU times are a few seconds on a VAXstation II/GPX. Also, OASYS was the first program to explicitly introduce hierarchy for analog integrated circuits [HAR_87]. This has several advantages: large design tasks are broken up in smaller, more manageable tasks; each high-level topology corresponds to many transistor-level circuits; knowledge of any subblock can be reused in many higher-level topologies and is much easier to change or update.

However, the program also shows some limitations. OASYS only provides some primitive form of optimization by iterating over a fixed number of plan steps. Due to the many heuristic decisions, the program is appropriate for providing initial design solutions only, which still have to be optimized by a local optimization program. In addition, all design plans are hard-codedly oriented towards some overall optimization goal, such as minimum power and area. If a different optimization goal is needed, for example minimum noise, then a different design plan has to be constructed. Finally, it is very difficult to include a new topology into the system. This requires entering the appropriate topology selection heuristics and the design plan for the new topology (and for all new subtopologies). For this, a lot of knowledge has to be collected in a fairly linear way and converted into program code by an expert designer. In practice, this shortcoming limits the program to a fixed set of topologies as well.

Therefore, experience was recently gained with a new implementation, OASYS-VM, which also contains a compiler for the newly developed analog description language OADL [OCH_89]. This language is aimed to facilitate the introduction of new topologies and knowledge into the system.

2.5.1.3. OPASYN and CADICS

OPASYN, developed at the University of California, Berkeley, U.S.A., is a compiler for CMOS opamps which accepts specifications and technology data and generates a correct layout of an optimized opamp [KOH_90]. It performs three actions: selection of an opamp topology, parametric optimization based on an analytic model of this topology, and generation of the layout in a macrocell place-and-route style. The principle block diagram of the OPASYN compiler is shown in Fig. 2.9.

OPASYN [KOH_90] currently contains 5 generic flat-level opamp topologies, which are assumed to be general enough to still cover a large range of applications. The rule-based selection program then selects one or more of these schematics by heuristically pruning a decision tree based on some key design specifications, such as the open-loop gain, and the technology data. All remaining schematics are sent to the optimization module, as the final selection is left to the user.

On each remaining topology a parametric optimization is then performed in order to optimize some user-defined performance objectives while meeting all specification constraints. For this optimization each opamp is characterized by an analytic model, which consists of a large set of nonlinear equations describing the circuit's behavior, some a priori design decisions and the list of independent design variables. This list has strongly been reduced by exploiting expert design knowledge. The analytic design equations have been derived by an experienced designer by means of first-order circuit analysis techniques and topology-specific approximations. Note that these equations also contain fitting parameters which allow to improve the accuracy by tuning with SPICE simulations. The optimization itself is based on a steepest-gradient method, starting from coarse-grid-sampled multiple starting points. The program always returns a solution: the circuit which best matches to the specifications (also if it does not meet all specifications).

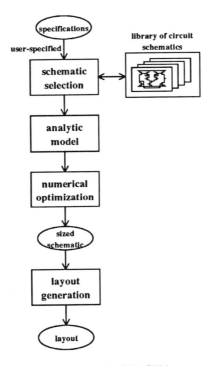

Fig. 2.9. Block diagram of the OPASYN opamp compiler.

After the optimization the layout is generated in a macrocell place-and-route style [KOH_90]. The placement is performed according to a predefined circuit-specific slicing tree. For the routing all signal nets are divided in different classes (e.g. sensitive nets, noisy nets...) which are routed subsequently.

Due to the optimization approach, OPASYN requires more time to size an operational amplifier. Still, it typically takes less than 5 minutes to compile an optimized opamp from specifications to layout on a VAX 8800. The program, however, also shows some limitations. First of all, it does not support any form of hierarchy and hence no modularity in the analytic models. Also, there is no failure recovery: the program just returns the best obtained solution (even if it does not meet all specifications) without suggesting any modifications to the specifications or to the topology. Finally, it requires an enormous effort to introduce a new schematic into the system, since for each schematic the corresponding analytic model and the required layout information have to be derived manually by an expert designer.

CADICS, on the other hand, also developed at the University of California, Berkeley, U.S.A., is a synthesis program for one particular type of A/D converters [JUS_90]: it sizes and generates the layout for one-bit-per-cycle cyclic A/D converters from a set of specifications and technology data. The program has two hierarchical levels. The specifications are translated (or mapped) from the converter level to the subblock level by an architecture-specific procedure. Each subblock is then synthesized by a customized routine. For example, for the synthesis of the opamps in the converter, OPASYN is invoked. The performance of the converter is then verified by means of a simple dedicated behavioral simulator built into the program. If needed, the synthesis is iterated. Then the layout of the converter is generated in an hierarchical way: the partitioning of the different subblocks is predefined but the layout of each subblock is generated in a macrocell place-and-route style. Extensions to other classes of A/D converters and thus a generalization of the synthesis framework are planned.

2.5.1.4. BLADES

BLADES, developed at AT&T Bell Labs, U.S.A., is an expert-systems-based analog circuit design environment [ELT_89]. It uses both formal (analytic) and intuitive knowledge for the circuit synthesis, it considers different levels of abstraction and aims to mimic the reasoning process of analog designers.

BLADES uses different levels of abstraction: it currently recognizes the component level, the subcircuit level and the circuit level, each described by an appropriate hardware description language. Starting from the input specifications, BLADES then synthesizes an appropriate analog circuit by building it up from functional subcircuit blocks and by translating the circuit specifications into subcircuit specifications. These subcircuits, also called design primitives, are then sized and assembled together. For example, the design primitives for an opamp are a differential pair, an active load, a current source, a push-pull stage, etc. Finally, the circuit is verified by numerical simulation. Most of the design knowledge is represented as if-then rules in a rule-based production system and are used to map the specifications from a higher level to a lower level.

Each subcircuit class is designed by a different software module, called a subcircuit design expert, which has its own knowledge of possible structures and their associated design equations. The present version of BLADES [ELT_89] is limited to a certain class of bipolar opamps and their subcircuits. Typical CPU times for an opamp are 3 to 6 seconds on a VAX 785. However, the program has no clear indication of the hierarchy used, nor how subcircuit interactions are managed. It has no clear failure recovery mechanism. It also has no form of optimization or iteration, and can thus only produce initial design solutions. Moreover, it seems difficult to extend the method to other functional blocks than opamps as well, or even to cover a wide range of performance specifications. The rules used in BLADES are hard to formulate in a general, context-independent way with a large scope. Finally, the program also has no automatic layout generation.

2.5.1.5. Short summary of other approaches

PROSAIC [BOW_85] applies expert-system techniques to synthesize CMOS operational amplifiers. The opamps are designed by decomposing them into stages. The schematics of these stages are selected from among a fixed set of alternatives stored in the database. During the subsequent interstage connection, the biasing and device sizes of each stage can be altered in order to obtain the required performance. However, the program has only limited hierarchy, lacks an adequate optimization and does not seem to be general enough to be extended to the design of other functional modules as well.

COARSE [HEI_88, HEI_89] sizes a limited set of fixed CMOS operational amplifiers in two phases. By means of some topology-specific equations and heuristics, it first generates an operating point for the opamp which satisfies all specifications. Starting from this initial solution, the circuit is then optimized with a minimax criterion by changing the DC operating point values and evaluating the circuit performance with a numerical simulator (but without DC analysis). The method, however, has no hierarchy and only seems to be applicable to operational amplifiers. COARSE has been incorporated as circuit design routine in the knowledge-based analog design system KADS, which is currently being developed as part of the NORSILC project [SAS_89].

AN_COM [BER_88a] designs a limited class of operational amplifiers by alternating topology selection and specification translation within a hierarchy of functional blocks. The layouts from the lowest-level blocks are then assembled to constitute the total layout of the opamp. In this way, AN_COM very much resembles OASYS [HAR_89b]. But there are some differences. AN_COM performs a macromodel-based simulation to verify any decomposition before going down in the hierarchy and it also translates geometrical specifications down the hierarchy. The decompositions in AN_COM are driven by knowledge stored in templates, which are more complicated than the design plans in OASYS. As a result, it is even more difficult to include new topologies into the system. AN_COM has not yet proven itself in many practical designs, and it does not seem to be applicable to higher-level circuits either. Recently, the same group has presented an algorithmic approach to compile high-performance medium-speed sampled-data systems from analog behavioral description down to layout [BER_89]. However, no practical results have been shown yet.

OAC [ONO_90] optimally sizes and generates a layout for CMOS operational amplifiers from performance specifications and technology data. The design is accomplished in two major steps. In the global design step, an opamp topology is selected from a limited library of transistor-level schematics and rough device sizing is performed. The topology is selected by choosing the circuit, which is most likely to meet the specifications, out of a set of existing design examples stored in the database. The design parameters in this circuit are then modified in order to improve its performance according to modification plans derived from built-in design knowledge. Then, in the detailed design step, both detailed device sizing and layout generation are carried out simultaneously. The circuit performance is optimized using a nonlinear optimization routine. At each iteration, the layout is then generated in a procedural way. Based on the extracted layout, the circuit is then simulated and the results are fed back to the optimization program. In this way, the sizing takes layout parasitics fully into account. It is, however, not clear whether this methodology can be extended to higher-level modules such as A/D converters, since the OAC has no hierarchy and thus a large number of design examples have to be stored into the database in order to cover a broad range of performance specifications. Also, the inclusion of new schematics and the corresponding design knowledge and modification plans into the system is not clear.

The above overview is not intended to be complete. It shows that many different approaches and programs exist for the automated design of analog building blocks. Most of these tools, however, are directed towards the design of operational amplifiers. Except for a class of comparators in OASYS [HAR_89b] and a class of A/D converters in CADICS [JUS_90], IDAC [DEG_89] today is the only program which can effectively design circuits from a broad range of different classes. There is, however, an increasing amount of research going on nowadays towards the automated design of higher-level analog modules, and practical results are to be expected in the near future. At the same time, several other tools have been published which attack the design of switched-capacitor filters [ASS_88, BAR_89, NEG_90, SIG_87, THE_82, THE_87, TRO_87, WAL_91]. This topic, however, is beyond the scope of this book.

The above overview clearly reflects the present state of the art in analog design automation. The discussion has also clearly shown that all existing systems have certain advantages and certain limitations. The major approaches are now compared in somewhat more detail in the following subsection.

2.5.2. Comparison of the existing analog design systems

The ASAIC system [GIE_90a], presented in section 2.3, and the major other analog design systems, presented in the previous subsection, ADAM [DEG_89], ACACIA [CAR_89], OPASYN [KOH_90] and BLADES [ELT_89], are compared in Table 2.2 with respect to the overall design strategy, and the degree and general approach of automation of the different design tasks. Note that in all of these systems, the analog circuits are designed in a custom way for each application.

	ASAIC	ADAM	ACACIA	OPASYN	BLADES
non-fixed-topology	yes	no	no	no	no
topology selection	automatic	manual	automatic	automatic	automatic
hierarchy	yes	limited	yes	no	yes
topology characterization	symbolic simulation	manual	manual	manual	manual
sizing method	optimization analytic models	synthesis + optimization	knowledge-based synthesis	optimization analytic models	knowledge-based synthesis
sizing algorithm	simulated annealing	formal procedure	expert system (design plans)	steepest gradient	expert system (rules)
optimization	yes	yes	limited	yes	no
failure recovery	yes	limited	yes	no	limited
layout generation	yes	yes	yes	yes	no

Table 2.2.Comparison of automated analogue design systems :
ASAIC, ADAM, ACACIA, OPASYN and BLADES.

From this table and from the previous discussion, it is clear that systems such as ADAM and OPASYN support no real hierarchy. ADAM is also the only system which does not provide an automatic topology selection, but leaves this task to the user. All the systems use some kind of analytic description (design equations) of the circuit to perform the sizing and all exploit expert design knowledge to some extent. The representation and the way of using this knowledge, however, is largely different in all methods. ASAIC and OPASYN for example are optimization-based, whereas the other systems are knowledge-based. The former systems thus require larger CPU times, but will usually result in closer-to-optimal designs. In addition, ASAIC is the only system up till now which automatically generates large part of the analytic information needed for sizing a circuit due to the use of symbolic simulation techniques. In this sense, ASAIC is a first approach towards creating an open analog design system which allows an easy inclusion of new topologies by the designer himself. It has to be mentioned though that similar extensions are being planned for ADAM and ACACIA as well. Next, automatic failure recovery is only provided in ASAIC and ACACIA. And BLADES does not provide any automatic layout generation. Finally, it has to be emphasized that at present ADAM is the only system which can design circuits from a large variety of different circuit classes such as voltage references, oscillators, etc.

Note also that ASAIC [GIE_90a] and OPASYN [KOH_90] both perform the sizing as an optimization of analytic circuit models. The important differences between both systems however are: OPASYN uses no hierarchy; OPASYN utilizes a local optimization algorithm whereas ASAIC uses simulated annealing; OPASYN has no backtracking mechanism to change specifications or the topology if the resulting design does not meet the specifications; the analytic models for OPASYN are derived manually.

The above systems can already design many practical circuits. As a suggestion for future research in this area, however, some major issues for developing an analog design system are now summarized below :
• The first problem of many existing tools is that **it is very difficult, if not practically impossible, for a designer to introduce a new schematic into the system**, due to the lack of any supporting tools and interfaces. As a result, these systems are limited to the fixed set of circuit topologies stored in the program's database (both if the system uses hierarchy or not). For analog integrated circuits, this is an essential shortcoming since many high-performance applications require minor or major modifications to existing schematics in order to meet the stringent performance requirements. In some experiment carried out in a telecommunication company for example, IDAC [DEG_87c], although it has a wide range of more than fifty schematics, turned out to cover only 25% of the company's analog design needs. Nevertheless, except for ASAIC [GIE_90a], none of the existing systems is able to automatically generate the appropriate design model and analytic equations for a new topology. And the manual derivation of design equations is a tedious and error-prone job, especially for large circuits. For IDAC for example, it has been published that the average development time for the inclusion of a new schematic ranges from 3 to 9 man-months of which about one-

half is spent for the analytic study and setup of the synthesis procedure [DEG_89]. Therefore, methodologies such as ASAIC which uses symbolic analysis techniques as an attempt to create an open design environment are becoming more and more important in the near future. Besides, similar extensions are now being planned for ADAM [DEG_90] and ACACIA [OCH_89] as well. This also justifies the detailed discussion of symbolic analysis techniques in the next three chapters.

- A related issue is the problem of the **acceptability of the automated design tools by the analog design community.** There are essentially two different types of users for an analog design system: system designers, who want to run the system automatically, and circuit designers, who want the design task to be interactive. Finding a good balance between both users groups however is not easy. The present analog design systems are often too complicated for system designers and not open enough for circuit designers, and are thus not acceptable for both kinds of designers. System designers want as little interaction at the circuit level as possible. For example, they certainly do not want to select any schematic themselves or to supply values for any degrees of freedom in the design as in IDAC [DEG_87c]. Circuit designers on the other hand want intensive interaction and want to be informed of any heuristic decisions and simplifications, internally taken by the system. They also want to include their own knowledge and design strategies into the system, which is at present nearly impossible due to the lack of any designer-friendly interface. Instead, the interaction with most existing programs is still performed at the level of program code and the designer really has to know all the program's implementation details. One solution to this problem of course is to target the design system to one group of users only. However, the authors of this book believe that the opposing requirements resulting from the two different groups of users, can be combined into one design system. This then must have a designer interface for explanation and knowledge/schematic introduction in addition to a fully automatic design mode. The ASAIC system [GIE_90a], presented in this chapter, is a first attempt to such a general and open design system for analog functional modules.

- Finally, there is also the discussion of **knowledge-based versus optimization-based design, which comes down to the trade-off of CPU time and performance quality of the resulting design.** Knowledge-based techniques heuristically decide on all degrees of freedom in the design. For example, OASYS [HAR_89a] distributes the overall gain in a two-stage opamp equally among both stages. (In IDAC, the decision on values for some degrees of freedom in the design is even left to the user [DEG_87c], which is a difficult task for system designers.) In this way, the design knowledge residing in the analog design community can be exploited, and an initial design solution can be obtained in a very fast way. However, the analog expert design knowledge is still far from formalized, and the problem of acquisition or extraction of the knowledge out of the designers has not yet been solved satisfactorily. In addition, the heuristically obtained initial design solution can be far from the global optimum solution (such that even local optimization cannot obtain the real optimum anymore). An alternative of course is the use of global optimization techniques such as in ASAIC [GIE_90b], which implicitly take care of all degrees of freedom and trade-offs in the design (such as the gain distribution between the two stages) and obtain a solution close to the global optimum. The major disadvantage of these methods of course is that they require larger CPU times.

2.6. Conclusions

In this chapter, an overview has been given of the state of the art in the field of automated design of analog functional modules (such as operational amplifiers, filters, comparators, A/D and D/A converters). The major tasks have been indicated and a classification of the existing design systems has been presented. This has then been illustrated by the detailed discussion of an example design system, ASAIC [GIE_90a].

ASAIC covers the whole design path from topology selection over analytic characterization and optimal sizing down to layout generation, and optimizes the circuit in a custom way for each application based on its analytic model. The different steps have then been illustrated for the design of an operational amplifier. Finally, ASAIC [GIE_90a] and other analog design systems presented in the literature, such as ADAM [DEG_89], ACACIA [HAR_89b], BLADES [ELT_89] and OPASYN [KOH_90], have been compared to reflect the state of the art in automated analog module design. The major feature of the ASAIC system then clearly is that it uses symbolic analysis and equation manipulation techniques as a solution to create an open design environment, which - in addition to a fully automatic mode - also offers additional interaction possibilities for expert circuit designers to insert new topologies and design knowledge into the design system. Since the AC characteristics for any new topology are automatically generated by means of the symbolic simulator ISAAC [GIE_89c], the effort to include a new topology into ASAIC is largely reduced. This now clearly motivates the detailed discussion of symbolic analysis techniques in the following chapters.

3   SYMBOLIC SIMULATION OF ANALOG INTEGRATED CIRCUITS

3.1. Introduction

There are two major reasons for the recent increasing interest in symbolic analysis techniques for analog integrated circuits. The first reason, as explained in the previous chapter, is that a symbolic simulator can be used to automatically generate large part of the analytic model of a circuit needed to size that circuit in an automated design system, and thus largely reduces the effort to introduce a new topology into the design system [GIE_90a]. The second reason arises from the fact that analog design is very knowledge-intensive and that the quality of the resulting design strongly relies on the insight and expertise of the designer himself. However, real analog experts are hard to find and even harder to keep, and most of the designers only master some limited part of the analog design field. In addition, it takes a too long time to train novice designers. Up till now, they have to acquire their own design knowledge from textbooks [ALL_87, GRA_84, GRE_84, GRE_86, LAK_91, TSI_85] and from long and tedious hand calculations and simulations. This is time-consuming and extremely difficult for performance-degrading effects such as noise and distortion.

For these reasons, the symbolic simulator ISAAC (Interactive Symbolic Analysis of Analog Circuits) has been developed [GIE_89c, SAN_89, WAL_89], 1) to provide insight to students and novice designers as well as to interactively assist more experienced designers, and 2) to automatically generate the analytic AC circuit models for the ASAIC analog design system [GIE_90a] presented in the previous chapter.

In this chapter, symbolic circuit analysis is discussed in general. A definition of symbolic simulation is provided, its advantages and application domains are pointed out, and an implementation of a symbolic simulator, ISAAC, is described. In the next section 3.2, first symbolic simulation is defined and the scope of the symbolic simulator ISAAC is depicted. Section 3.3 then describes symbolic circuit analysis and its advantages and differences with numerical methods. The main applications of symbolic simulation in analog circuit design are discussed. Finally, in section 3.4, a general description of the structure and functional capabilities of the ISAAC program is given, which is illustrated with several examples.

3.2. Definition and scope of symbolic simulation

Generally stated, a physical system can be described quantitatively and qualitatively, as shown in Fig. 3.1. Both formalisms are abstractions of the actual behavior of the system, and provide complementary insight into the behavior of this system. In **qualitative simulation**, the variables in the system analysis can only take a limited range of qualitative values e.g. {increase, decrease, no_change} and a perturbation on the input nodes is propagated to the output node by a causal reasoning process [DEK_84]. In **quantitative simulation**, the variables in the system analysis can continuously take all real values as a function of time and the response of the system to a given excitation can be calculated by means of numerical or analytical (symbolic) techniques. *Numerical techniques* discretize the time or frequency axis and numerically calculate the output response at every time/frequency point by means of some numerical algorithm. *Analytical techniques* try to find a closed-form expression for the system's response (in which the system parameters can be represented by a symbol or by their numerical value). This symbolic analysis is the topic of this book. Note however that **numerical, symbolic and qualitative simulation are all useful complementary techniques, and that they provide different information about the same circuit.**

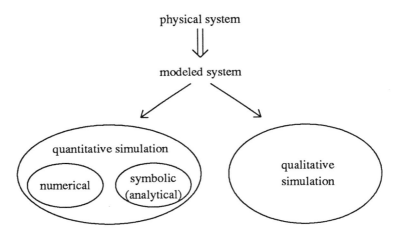

Fig. 3.1. Quantitative and qualitative simulation as abstractions of the actual behavior of a system.

Up till now, the scope of symbolic analysis is restricted to the analysis of both linear and weakly nonlinear lumped time-invariant circuits in the complex frequency domain, i.e. the Laplace domain (*s*-domain) for time-continuous circuits and the *z*-domain for time-discrete circuits. Symbolic time-domain analysis is not yet feasible up till now. The behavior of such a linear (or weakly nonlinear) lumped time-invariant circuit can then for zero initial conditions completely be described by symbolic (analytic) expressions for the transfer function or in general any network function of the circuit.

We can now define a symbolic simulator. A **symbolic simulator** is a computer program that performs a symbolic analysis on a model of the circuit under analysis for specific inputs and outputs, in a numerical-simulator-like user environment.

All concepts and techniques introduced in this book will be illustrated by one successful example implementation of a symbolic simulator, called ISAAC (Interactive Symbolic Analysis of Analog Circuits) [GIE_89c, SAN_89] and developed at the Katholieke Universiteit Leuven, Leuven, Belgium. ISAAC analyzes lumped, linear, time-invariant circuits in the complex frequency domain and returns analytic expressions for any network function with the complex frequency and (all or part of) the circuit elements represented by symbols. The program derives all AC characteristics for any analog integrated circuit : both time-continuous and switched-capacitor, CMOS, JFET and bipolar, and returns analytic formulas for transfer functions, CMRR, PSRR, impedances, noise... Time-continuous circuits are analyzed in the s-domain (Laplace domain), time-discrete circuits in the z-domain. Nonlinear circuits are linearized around their DC operating point, yielding the small-signal equivalent circuit. An extension to the symbolic distortion analysis of weakly nonlinear circuits is also provided [WAM_90], and will be discussed in chapter 5. The following example illustrates the idea of symbolic simulation.

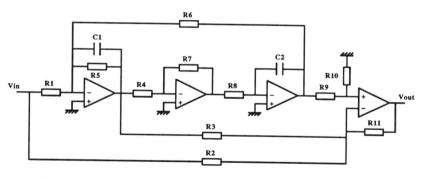

Fig. 3.2. Active RC filter to illustrate the principle of symbolic simulation.

Example
Consider the active RC filter of Fig. 3.2. The symbolic transfer function (with ideal operational amplifiers) is given by :

$$
\frac{\begin{array}{l} - \text{G4 G8 (G1 G2 G9 + G1 G3 G9 + G1 G9 G11 + G2 G6 G9} \\ \qquad \text{+ G2 G6 G10)} \\ + \text{S G7 C2 (G1 G3 G9 + G1 G3 G10 - G2 G5 G9 - G2 G5 G10)} \\ + \text{S}^2 \text{ (-1) G2 G7 C1 C2 (G9 + G10)} \end{array}}{\text{G11 (G9 + G10) (G4 G6 G8 + S G5 G7 C2 + S}^2\text{ G7 C1 C2)}}
\tag{3.1}
$$

Note that each resistor RX has been converted into the corresponding conductance GX.　　　　　　　　　　　　　　　　　　　　　　　　　　　　　◆

In general, the network functions for lumped, linear, time-invariant circuits are **rational functions** (ratio of two polynomials) in the complex frequency variable x (s or z) [ADB_80] :

$$
H(x) = \frac{N(x)}{D(x)} = \frac{\sum_i x^i \cdot a_i(p_1, \ldots, p_m)}{\sum_i x^i \cdot b_i(p_1, \ldots, p_m)}
\tag{3.2}
$$

The coefficients $a_i(\,\ldots\,)$ and $b_i(\,\ldots\,)$ of each power of x for both the numerator and denominator polynomial are symbolic polynomial functions in the circuit elements p_j, that are represented by a symbol instead of a numerical value.

This leads us to the **definition of the degrees of symbolization**. Depending on whether all, some, or none of the circuit elements are represented by symbols, three types of symbolic network functions can be distinguished [CHU_75] :

<u>Type 1</u> : fully symbolic network function

$$
\text{Example :} \quad \frac{V_{out}}{V_{in}} = \frac{1}{1 + RC\ s + LC\ s^2}
\tag{3.3}
$$

<u>Type 2</u> : partially symbolic or mixed symbolic-numerical network function

$$
\text{Example :} \quad \frac{V_{out}}{V_{in}} = \frac{1}{1 + 5.10{-}11\ R\ s + 3.10{-}14\ s^2}
\tag{3.4}
$$

$$
(C = 50\ pF, \quad L = 0.6\ mH)
$$

<u>Type 3</u> : rational function of s (or z) with numerical coefficients

$$
\text{Example :} \quad \frac{V_{out}}{V_{in}} = \frac{1}{1 + 5.10{-}6\ s + 3.10{-}14\ s^2}
\tag{3.5}
$$

$$(R = 100 \ k\Omega, \ C = 50 \ pF, \ L = 0.6 \ mH)$$

All three types are provided in ISAAC.

In this section, symbolic circuit simulation has been defined and the scope of an implemented simulator, ISAAC, has been depicted. In the following section, the applications of symbolic analysis in analog design are discussed.

3.3. Applications of symbolic analysis in analog design

In this section, it is shown how symbolic simulation forms an essential complement to numerical simulation. A symbolic simulator gives the designer a different perspective than that provided by numerical simulators, which is especially useful to gain insight into a circuit's behavior [GIE_89c]. In addition, a symbolic simulator in combination with a versatile plot program constitute a convenient environment for interactive circuit improvement and exploration. Symbolic expressions are also useful for the repetitive evaluation of network functions [LIN_73b]. Moreover, a symbolic simulator can be used to automatically generate analytic circuit models for automated analog circuit dimensioning, thereby strongly reducing the effort and time needed to include new topologies in the design system [GIE_90b]. Finally, it enables the behavioral simulation of complex circuits.

3.3.1. Insight into circuit behavior: symbolic versus numerical simulation

Nowadays, many numerical circuit simulators (such as SPICE [VLA_80b] and its commercial derivatives, SWAP [SWA_83]) exist. Although very efficient, they only return a series of numbers, in tabulated or plotted form. The only conclusion that can be drawn from these numbers is whether the circuit meets the specified functional behavior, for the topology description and the technological and environmental parameters provided in the input file. No indication is given of which circuit elements essentially determine the observed performance. No potential problems are pointed out. No solutions are suggested when the circuit does not meet the specifications. To scan performance trade-offs and to check the influence of parameter variations, many simulations have to be carried out. Yet the results still cannot be extrapolated safely. This is especially the case for second-order phenomena such as the power-supply rejection ratio (PSRR), which strongly depends on component mismatches and layout overlap capacitances. It can be concluded that numerical circuit simulators are mainly useful for functional (nominal) verification, during or after the design of the circuit.

On the other hand, a symbolic simulator, such as ISAAC, can greatly improve the designer's insight into a circuit's behavior. A symbolic simulator replaces the long and tedious hand calculations of the analog designer, especially for second-order effects, such as distortion and PSRR. It returns first-time correct analytic expressions in a much shorter time and for more complex characteristics and circuits than would be possible by hand. The resulting expressions are valid whatever the element values are. The same expressions hold when the parameter values change (assuming that all transistor small-signal models remain valid). By examination of the resulting expressions, especially when they are simplified to retain the dominant terms only, insight can be gained into the circuit's behavior [GIE_89c]. The analytic expressions clearly indicate the fundamental design variables. They show performance trade-offs and sensitivities to parameter variations and are suitable for pole-zero extraction, for sensitivity and tolerance analysis [SIN_77] and even for fault diagnosis [LIB_88].

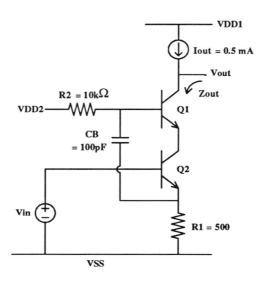

Fig. 3.3. Bipolar cascode stage with bootstrap capacitor to increase the output impedance at higher frequencies.

Since a symbolic analysis program provides insight into the behavior of analog circuits, it is more useful for the instruction of students and novice designers than a numerical simulator such as SPICE [VLA_80b]. Circuits can be explored without any hand calculation in a fast and interactive way. In addition, a symbolic simulator is also a valuable aid for experienced designers to check their own intuitive knowledge. Indeed, experiences with ISAAC have shown that even expert designers are sometimes wrong in writing down simplified formulas for circuit characteristics. This is especially the case for nonconventional circuit configurations, second-order characteristics, circuits with internal feedback, etc., where designers often make wrong assumptions or inappropriate simplifications. It can be concluded that a symbolic simulator is a valuable analog design aid, which forms an essential complement to numerical simulators [GIE_89c, SAN_89]. The differences between symbolic and numerical simulation are now illustrated in the following example [GIE_89c].

Example

Consider the bipolar cascode stage, depicted in Fig. 3.3. A bootstrap capacitor C_B has been added, which is expected to increase the output impedance at higher frequencies. A SPICE result of the output impedance versus the frequency for a C_B value of 100 pF is shown in Fig. 3.4. The transistors are modeled by r_π, g_m and g_o only. The plot confirms the expected impedance improvement. However, no indication is given which elements determine the impedance levels at low and high frequencies or the two break points. Even an experienced designer has difficulties to formally predict the influence of C_B on the output impedance, although it is a simple two-transistor circuit.

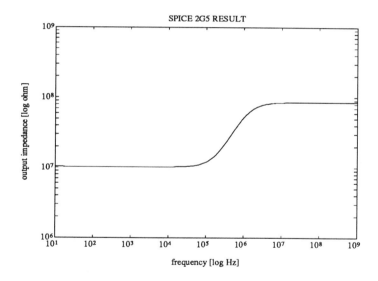

Fig. 3.4. SPICE result for the output impedance of the bootstraped bipolar cascode stage.

The same circuit is now analyzed with the symbolic simulator ISAAC. The input file to the program is shown below. It is stated in the well-known SPICE format [VLA_80b].

```
BOOTCAS
*
VDD1 1 0 DC 10
VDD2 2 0 DC 5
Q1 4 3 5 0 NPN
Q2 5 7 6 0 NPN
R1 6 0 500
R2 2 3 10K
CB 3 6 100P
VIN 7 0
IOUT 1 4 DC 0.5M
*
.MODEL NPN NPN VAF=50
```

The complete symbolic expression for the output impedance is derived by ISAAC as :

$$
\begin{aligned}
&\text{G2 GM.Q1 GM.Q2 + G1 G2 GM.Q1 + G2 GM.Q2 G}\pi\text{.Q1}\\
&\quad\text{+ G2 GM.Q1 G}\pi\text{.Q2 + GM.Q2 GO.Q1 G}\pi\text{.Q1 + G1 G2 G}\pi\text{.Q1}\\
&\quad\text{+ G2 GM.Q2 GO.Q1 + G2 GM.Q1 GO.Q2 + G1 GO.Q2 G}\pi\text{.Q1}\\
&\quad\text{+ G2 G}\pi\text{.Q1 G}\pi\text{.Q2 + G1 GO.Q1 G}\pi\text{.Q1 + G1 G2 GO.Q2}\\
&\quad\text{+ G1 G2 GO.Q1 + GO.Q2 G}\pi\text{.Q1 G}\pi\text{.Q2 + GO.Q1 G}\pi\text{.Q1 G}\pi\text{.Q2}\\
&\quad\text{+ G2 GO.Q2 G}\pi\text{.Q1 + G2 GO.Q1 G}\pi\text{.Q2 + G2 GO.Q2 G}\pi\text{.Q2}\\
&\quad\text{+ GO.Q1 GO.Q2 G}\pi\text{.Q1 + G2 GO.Q1 GO.Q2}\\
&\text{+ S CB (GM.Q1 GM.Q2 + G1 GM.Q1 + GM.Q1 G}\pi\text{.Q2 + G2 GM.Q1}\\
&\qquad\text{+ G1 G}\pi\text{.Q1 + GM.Q2 GO.Q1 + G}\pi\text{.Q1 G}\pi\text{.Q2 + G1 GO.Q2}\\
&\qquad\text{+ G2 G}\pi\text{.Q1 + G1 GO.Q1 + GO.Q1 G}\pi\text{.Q2 + GO.Q2 G}\pi\text{.Q2}\\
&\qquad\text{+ GO.Q1 G}\pi\text{.Q1 + G2 GO.Q1 + G2 GO.Q2 + GO.Q1 GO.Q2)}\\[8pt]
\hline\\[-4pt]
&\text{GO.Q1 (G2 GM.Q2 G}\pi\text{.Q1 + G1 G2 G}\pi\text{.Q1 + G2 G}\pi\text{.Q1 G}\pi\text{.Q2}\\
&\qquad\text{+ G1 GO.Q2 G}\pi\text{.Q1 + G1 G2 GO.Q2 + GO.Q2 G}\pi\text{.Q1 G}\pi\text{.Q2}\\
&\qquad\text{+ G2 GO.Q2 G}\pi\text{.Q2 + G2 GO.Q2 G}\pi\text{.Q1}\\
&\qquad\text{+ S CB (G1 G}\pi\text{.Q1 + G}\pi\text{.Q1 G}\pi\text{.Q2 + G2 G}\pi\text{.Q1 + G1 GO.Q2}\\
&\qquad\qquad\text{+ GO.Q2 G}\pi\text{.Q2 + G2 GO.Q2)))}
\end{aligned}
$$

(3.6)

Note the dot notation which is used in the program to denote the small-signal components of all transistors. The above expression is already lengthy for such a simple circuit. For a bipolar transistor, however, it usually holds that GM >> Gπ >> GO. Exploiting this relative-size information, the expressions can be simplified, thereby introducing some error. The error definition and the simplification algorithm will be described in the next chapter. The simplified output impedance for a 10% error is given by :

$$
\frac{\text{GM.Q1 (GM.Q2 + G1) (G2 + S CB)}}{\text{GO.Q1 G}\pi\text{.Q1 (G2 (GM.Q2 + G1) + S CB (G1 + G}\pi\text{.Q2))}}
$$

(3.7)

The approximation can be carried out further on, yielding simpler and simpler formulas, at the expense of more and more error. If a 25% error is allowed, the output impedance is given by :

$$\frac{GM.Q1 \; GM.Q2 \; (G2 + S \; CB)}{GO.Q1 \; G\pi.Q1 \; (G2 \; GM.Q2 + S \; CB \; G1)} \tag{3.8}$$

From (3.8), the impedance levels and the pole and zero can easily be retrieved :

$$Z_{out}(low \; f) \; = \; \frac{GM.Q1}{G\pi.Q1 \; GO.Q1} \tag{3.9}$$

$$z \; = \; - \; \frac{G2}{CB} \tag{3.10}$$

$$p \; = \; - \; \frac{GM.Q2 \; G2}{G1 \; CB} \; = \; \frac{GM.Q2}{G1} \; z \tag{3.11}$$

$$Z_{out}(high \; f) \; = \; \frac{GM.Q1 \; GM.Q2}{G\pi.Q1 \; GO.Q1 \; G1} \; = \; \frac{GM.Q2}{G1} \; Z_{out}(low \; f) \tag{3.12}$$

At lower frequencies, this stage actually does not behave as a cascode stage due to the base currents of the bipolar transistors. The "external" impedance at the emitter of Q1 is up-transformed to the output with a factor (GM.Q1 RO.Q1). However, this impedance is determined by Rπ.Q1 (in series with the smaller R2) and not by R1 * (GM.Q2 RO.Q2), as one would expect. This results in the output impedance given by (3.9). On the other hand, at higher frequencies, the bootstrap capacitor CB connects Rπ.Q1 and RO.Q2 in parallel and the cascode formula holds again. The output impedance is now given by (3.12), which is a factor (R1 GM.Q2) higher than at lower frequencies. This behavior is summarized in Fig. 3.5. The symbolic simulator clearly gives more insight into the circuit than SPICE (Fig. 3.4) does. Of course, at higher frequencies, the output impedance decreases again due to other capacitances in the circuit, which have not been included in the SPICE and ISAAC simulations. ♦

3.3.2. Interactive circuit improvement

A symbolic simulator also enables interactive circuit improvement [GIE_89c]. The symbolic expressions immediately translate topology changes into performance changes. For example, the influence of adding an extra component can be examined in the expressions of the new circuit. This is especially interesting in combination with a versatile plot program. The expressions can then be drawn and explored graphically. In this way, the interactive synthesis of new, high-performance circuits becomes feasible.

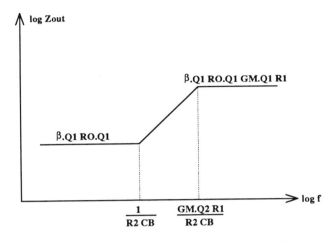

Fig. 3.5. ISAAC result for the output impedance of the bootstraped bipolar cascode stage.

This leads us to the proposition of an interactive design and exploration environment for both the experienced and the inexperienced analog circuit designer. This environment comprises a schematic editor (with automatic generation of simulator input files), a symbolic and numerical simulator, a versatile plot program (Bode plots, root loci, 3-dimensional graphics...), and possibly some optimization routines, all integrated in a common database.

Moreover, in [KON_88], a method is described for the automatic generation of new circuit structures with the combined aid of a symbolic simulator and the PROLOG language. A PROLOG program exhaustively generates all possible topologies with a predefined set of elements (for example one operational amplifier and three capacitors). The switched-capacitor circuit symbolic simulator SCYMBAL then derives a symbolic characteristic function for each structure. A PROLOG routine finally evaluates this function to check whether it fulfils all requirements (for example a stray-insensitive switched-capacitor integrator function with low opamp gain sensitivity).

3.3.3. Repetitive calculation of formulas

Symbolic formulas are also useful if a circuit has to be evaluated repeatedly for several values of the components or the frequency [LIN_73b]. It takes less time to evaluate the same analytic expression with one or a few parameters or the frequency varying over some range than to perform a full AC analysis each time. Much CPU time can be gained by exploiting the technique of compiled-code simulation: the behavior of an analog circuit is compiled into symbolic expressions which are then evaluated for particular values of the circuit and input parameters. This can be useful in tolerance (for example Monte Carlo) or large-scale sensitivity analysis, in design centering or yield estimation techniques [DIR_90].

Also, many second-order effects, such as the offset voltage and the PSRR, are essentially determined by device mismatches and tolerances. Numerical simulations for nominal design parameters can by far underestimate these second-order effects, whereas in ISAAC mismatches are explicitly taken into account [GIE_89c]. Hence, the influence of tolerances and mismatches can be seen right from the symbolic expressions.

In this context, it has to be mentioned that during the last few years, symbolic analysis is also more and more applied to digital (logic) circuits as well [BOL_89a, BOL_89b, BOS_89, BRY_87, VOS_88, VOS_89]. For these applications, the circuit is represented by a system of Boolean equations. The solution of this system provides for every circuit node a symbolic description of the possible conducting paths that influence this node [BOL_89b]. In this way, the logic function performed by a circuit can be extracted in symbolic form. This symbolic information can then be applied to solve numerous problems in the area of digital circuit design. For example, a symbolic logic simulation performed by evaluating the extracted formulas of a circuit for input patterns consisting of Boolean symbolic variables is equivalent to exhaustively simulating the circuit with traditional logic simulators, but requires a much shorter CPU time [BOL_89b]. Also, the Boolean description of a circuit can be combined with formal proof techniques for the functional verification of the circuit. The symbolic information can also be exploited during electrical and timing diagnosis of MOS circuits [BOL_89a].

3.3.4. Analytic model generation for automatic circuit dimensioning

An important application of symbolic simulation is to integrate it as a CAD module into an automated analog design system. As already discussed in the previous chapter, for the automation of analog design, it is important that analog designers are able to easily introduce new circuit schematics and to easily add their design knowledge to the synthesis system [GIE_90a]. This problem is partially solved by conceiving the design system as a flexible framework with a friendly interface to ease the process of adding design knowledge by the designer. IDAC will evolve towards a framework in which experienced designers can add their circuit and their own knowledge in the form of a tableau description [DEG_89, DEG_90]. This tableau contains the net list and the contributions of devices to several performances, and will be compiled into the appropriate IDAC code afterwards. Similarly, OASYS-VM will facilitate the inclusion of detailed circuit design knowledge in the form of design plans and design styles into the OASYS framework [OCH_89].

However, two problems still remain then with such an open design framework. The first one is the lack of structured analog design knowledge. The expertise residing in analog designers is rather intuitive in nature and hence difficult to formalize. Possibly, knowledge extraction techniques emerged from artificial intelligence research can be applied here. The second problem is that the analytic formulas for a new circuit still have to be derived manually. To overcome this, symbolic analysis techniques are most appropriate. Symbolic simulators such as ISAAC, which can characterize and set up analytic models for new circuits, are what is needed in future analog design expert systems development [SAS_89]. As symbolic circuit analysis techniques mature, the effort to introduce new circuits into analog design systems may become significantly smaller [KOH_90]. Therefore, in the ASAIC system, which has been described in the previous chapter, the analytic circuit models are to a large extent generated automatically by calls to the symbolic simulator ISAAC [GIE_90b]. In this way, ISAAC forms the cornerstone of the ASAIC non-fixed-fixed-topology analog module design system. Similarly, in the future, the tableau for IDAC will be generated automatically, starting from a schematic entry, by the symbolic analysis program SYNAP [DEG_90, SED_88].

Also, the use of simplified symbolic formulas can strongly speed up the numerous evaluations of the objective function and boundary conditions during the optimization of large circuits [GIE_89a, GIE_90b]. In addition, manual or automatic inspection of the analytic circuit models possibly yields rules, which can further reduce the design search space.

3.3.5. Simulation of complex circuits

In spite of the increasing efficiency of the algorithms [KUN_86a], the maximal size of a numerically simulated circuit is still limited. One cannot simulate entire, eventually mixed analog-digital systems. A solution to this problem which still shows acceptable CPU times, is high-level simulation, in which all building blocks are characterized by a behavioral model or a macromodel. At present, however, the derivation of these behavioral or macromodels is carried out manually. A symbolic simulator can be used to automatically generate parametrized functional models for building blocks such as amplifiers, filters, etc. Indeed, the characteristics of a particular circuit can be derived and compiled symbolically, and then evaluated after all circuit parameters have been determined. Alternatively, if the circuit has already been designed, the values of all parameters can be used to generate exact expressions in the symbolic frequency variable s or z only.

Also, the combination of symbolic and numerical methods could be useful for the simulation of large circuits with waveform relaxation techniques [RUE_86]. The analytic solution of a symbolic simulation program (which does not have to be exact at all) can be used as a first solution for the transient analysis in a waveform relaxation program. This will reduce the total simulation time, as the first iteration is often the most difficult one. Moreover, for this application, the numerical values of all elements are known and can be used in the symbolic simulator, resulting in network functions of type 3. This strongly reduces the CPU time required for the symbolic simulation and, in combination with hierarchical techniques, allows the symbolic analysis of large circuits as well. A similar approach can be followed for the electro-thermal simulation of integrated circuits [PET_90].

Conclusion

In this section, the applications of symbolic simulation in analog design and analog design automation have been discussed. First, it has been shown how symbolic simulation forms an essential complement to numerical simulation for providing insight into the behavior of a circuit. Secondly, a symbolic simulator in combination with a versatile plot program constitute a convenient design environment for interactive circuit improvement and exploration. Thirdly, symbolic expressions are also useful for the efficient repetitive evaluation of network functions. Fourthly, a symbolic simulator can be used to automatically generate analytic circuit models for automated analog circuit dimensioning, thereby strongly reducing the effort and time needed to include new topologies in the design system. Finally, it enables the behavioral simulation of complex circuits.

With these applications in mind, the symbolic simulator ISAAC has been developed. It will be described in the next section, where the program's functionality and flowchart are presented.

3.4. General description of the ISAAC program

With the applications of last section in mind, the symbolic simulator ISAAC [GIE_89c, SAN_89] has been developed. This program will be used in the remainder of this book as a representative example of a symbolic simulator for analog circuits, to illustrate the functional and algorithmic aspects of symbolic simulation. In this section, ISAAC's functionality and general flowchart are described. As compared to existing symbolic simulators, ISAAC is the first one which is fully dedicated and implemented towards analog (integrated) circuits and it is also the first one which offers a built-in approximation [SAN_89]. At the same time, ISAAC combines a very broad functionality with efficient internal algorithms [GIE_89c]. The program is then also extensively being used in several universities and in industry.

Because of the symbol processing and the ease of programming (fast prototyping), the program has been written in COMMON LISP [STE_84]. Therefore, ISAAC runs on any platform with a COMMON LISP compiler. In section 3.4.1, the analysis modes of the program are described. Section 3.4.2 then presents the available network elements. Finally, in section 3.4.3 the program flowchart and an overview of the most important program modules are given.

3.4.1. Analysis modes

ISAAC integrates the analysis of both time-continuous and time-discrete circuits in one and the same program. It currently provides the user with four analysis modes, as depicted in Fig. 3.6 [GIE_89c].

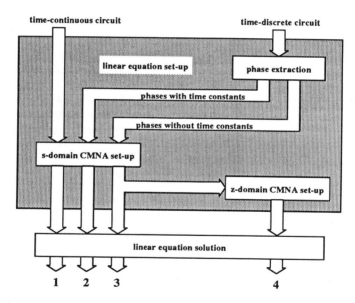

Fig. 3.6. Analysis modes of the symbolic simulator ISAAC.

I. Time-continuous circuits, such as opamps and active RC filters, are analyzed in the s-domain (path 1 in Fig. 3.6). A typical example is the gain of the nMOS single-transistor amplifier, shown in Fig. 3.7 :

```
      - GS (GM.M1 - S CGD.M1)
```
_____   (3.13)
```
      GS (GL + GO.M1)
  + S  (GM.M1 CGD.M1 + GL CGS.M1 + GL CGD.M1
        + GO.M1 CGS.M1 + GO.M1 CGD.M1 + GS CL
        + GS CBD.M1 + GS CGD.M1)
  + S²  (CL CGD.M1 + CGD.M1 CBD.M1 + CGS.M1 CGD.M1
        + CL CGS.M1 + CBD.M1 CGS.M1)
```

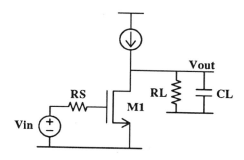

Fig. 3.7. nMOS single-transistor amplifier.

II. Time-discrete circuits with periodically operated switches, such as switched-capacitor filters, are analyzed with zero time constants in the z-domain (path 4 in Fig. 3.6). A typical example is the z-domain voltage gain from a phase-1 sampled-and-held input signal to the output sampled at phase 1 for the Fleischer-Laker biquad [FLE_79], shown in Fig. 3.8 :

```
      D K + D J - A L - A H
  + Z  (- 2 D K + A L + A G - D J - D I)
  + Z² D (K + I)
```
_____   (3.14)
```
  A E - D B + Z (- A E + 2 D B - A C + D F)
  + Z² (-1) D (F + B)
```

III. In the presence of nonzero time constants, the analysis of a time-discrete circuit requires a combination of s- and z-domain analyzes. A typical example is the influence of the finite opamp gain-bandwidth (an s-domain effect) on a switched-capacitor filter characteristic (in the z-domain). This mode, however, is not provided in the program, as it requires major modifications to the program's routines.

IV. Each phase of a time-discrete circuit can be analyzed separately in the s-domain with nonzero time constants (path 2 in Fig. 3.6) or with zero time constants (path 3 in Fig. 3.6). A phase is defined as a periodically returning time-interval in which the clocks of a time-discrete circuit have a certain logical level. Hence, every transition of one of the clocks in one general clock period creates a new phase.

The term phase is also used for the corresponding circuit with all switches closed or open, depending on the logical level of their clock in that phase. As an example of an s-domain analysis without time constants, the load impedance of the first opamp in the Fleischer-Laker biquad (Fig. 3.8) during phase 1 is given by :

$$\frac{S \; D \; (C + E + G + H + L)}{C + D + E + G + H + L} \tag{3.15}$$

which is capacitor D in series with capacitors C, E, G, H, and L in parallel. For an analysis with nonzero time constants, the on-resistance of the switches is symbolically taken into account. The results are quite cumbersome though.

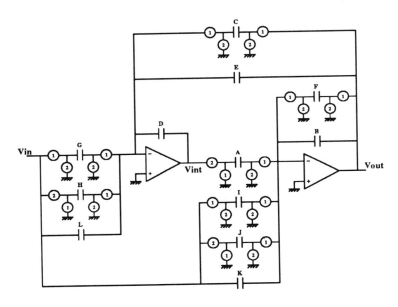

Fig. 3.8. Switched-capacitor biquad of Fleischer and Laker.

3.4.2. Circuit elements

Fig. 3.9 shows the primitive elements available in the ISAAC program [GIE_89c]. These include all basic elements from circuit theory: independent voltage and current source, resistor, capacitor, inductor, current-controlled current and voltage source, voltage-controlled current and voltage source, nullator and norator. Two measuring elements have been added: current meter and voltage meter. For time-discrete circuits a switch, a short and a clock have been introduced. For filters an ideal opamp has been added (in fact a combination of a nullator and a norator).

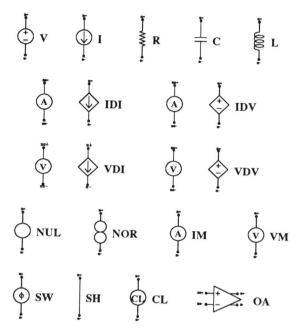

Fig. 3.9. Primitive elements available in ISAAC.

The user can define other nonprimitive elements (subcircuits) in terms of primitive elements (or other subcircuits). During the analyzes, these subcircuits are then recursively expanded into their primitive components. Because of this hierarchical feature, the **dot notation** has been introduced. The symbol which represents the child (component) X of a subcircuit Y is denoted by X . Y . If Y is also the child of another subcircuit Z, then X is denoted by X . Y . Z, and so forth. Note that diodes and transistors are predefined subcircuits for ISAAC, with the small-signal components as their children. Hence, the transconductance GM of transistor M8 is represented by the symbol GM . M8 in the ISAAC expressions.

3.4.3. Program structure

In this section, the structure of the ISAAC program is described. Fig. 3.10 shows the program's flowchart [GIE_89c]. The main modules are the input stage (with the read-in and the expansion of the circuit and the topology check), the analysis selection, the circuit transformations, the set-up and solution of the linear circuit equations, the heuristic approximation of the expressions and the output stage (with the output of the symbolic expressions and possibly a numerical evaluation). These main modules are now described shortly one after another. Note that this section focuses on the functionality of the program. Algorithmic details are provided in the next chapter.

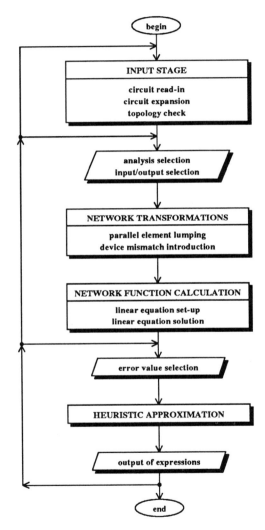

Fig. 3.10. Flowchart of the symbolic simulator ISAAC.

3.4.3.1. Input stage

At the input of the program, the circuit is read in, expanded and stored in the database. Next, a topology check is performed.

A. Circuit read-in

At the input stage, the circuit topology is read in by means of a conventional SPICE [VLA_80b] or SWAP [SWA_83] file for time-continuous or time-discrete circuits, respectively. In this way, compatibility with existing numerical simulators is established. Designers can use the same files and format as they did before. The only difference is that values do not necessarily have to be supplied and that no control statements have to be added, since the selection of the analysis type, the input source and the circuit output occur interactively during the ISAAC run. The input file only describes the circuit topology. Alternatively, the circuit can also be described in the VERA format [KOS_89] by means of a network description file and a type description file. Regardless of the input format, the circuit is then transformed into the internal datastructure, which is fully VERA-compatible.

B. Circuit expansion

Next, the circuit is expanded. Subcircuits are expanded into their primitive components. Nonlinear devices such as diodes and transistors are replaced by their small-signal model. These models are controlled by the user through the model statements (or the type description). As a general rule, the small-signal models are topologically the same as would be used in a SPICE run with the same input file [VLA_80b]. All SPICE defaults remain valid in ISAAC. Commonly used small-signal models are shown in Fig. 3.11 for (a) a diode, (b) a bipolar transistor, (c) a JFET transistor and (d) a MOS transistor. These default models are quite general and independent of the technology process. The MOS model for instance is independent of the region of operation of the transistor. It is topologically valid both in the linear and saturation region, both in strong and weak inversion. Of course, the physical quantity represented by a particular symbol, its relation to geometrical aspects and its magnitude do depend on the region of operation.

On the other hand, the more complex the small-signal expansion model of semiconductor devices becomes, the more terms the final expression contains. Most of these terms give only a minor contribution to the result, but they strongly increase the analysis time and storage requirements. This reveals the importance of the user-definable model facility. The user can select the appropriate model according to the application area he has in mind. High-frequency applications for example require a more sophisticated model than voice- or audio-band applications. So for the latter, the number of model parameters and the analysis time can be reduced without any loss of accuracy. Also, the small-signal model can differ from transistor to transistor.

This allows unimportant parasitic capacitances to be dropped, which strongly speeds up the analysis.

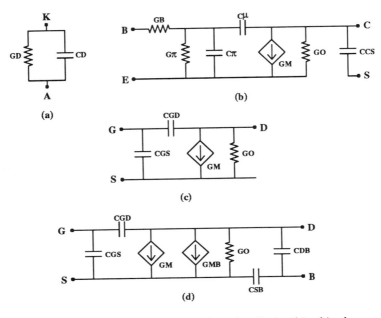

Fig. 3.11. Small-signal models for (a) a diode, (b) a bipolar transistor, (c) a JFET transistor, and (d) a MOS transistor.

C. Topology check

On the expanded network a topology check is performed. This prevents electrical and algorithmic problems during the network function calculation afterwards and signals potential problems to the user. Such problems may be caused by inconsistent or indeterminate circuit equations, redundant or superfluous circuit elements, typing errors in the input file, etc.

For ISAAC, three checks are sufficient to prevent virtually all mathematical and software errors for a time-continuous circuit. First, it is checked if the number of nullators is equal to the number of norators. Otherwise, the circuit equations would be overconstrained or underconstrained. Next, unallowed loops of zero-impedance elements are detected. Finally, unallowed cut sets of infinite-impedance elements are traced. For time-discrete circuits, these three tests are applied to the global circuit and to every individual phase as well [WAL_91]. The loop test and the cut-set test are now described in more detail.

- **Loop test**

Unallowed loops of zero-impedance elements (short, current meter, independent voltage source, current-controlled and voltage-controlled voltage source, nullator and opamp input) are detected before any analysis is done. Indeed, each of these elements fixes the element's branch voltage. Hence, in a loop of n such elements, the n branch voltages are determined. Still, Kirchhoff's voltage law has to be satisfied. As a result, the circuit can be inconsistent, undetermined, redundant or meaningless, unless the loop consists of shorts only. For example, if the loop contains a current meter, the current through this meter is undetermined and thus also the output signal if it depends on this current. If the output signal does not depend on this current, the current meter is redundant.

To detect a loop in the circuit, a graph search method is applied. All zero-impedance elements are replaced by an undirected branch connecting their n^+ and n^- node. The other elements are removed. The resulting undirected graph is decomposed into maximally connected subgraphs. In every subgraph, it is a priori checked whether any loop can be present. A subgraph cannot contain a loop if the number of branches is smaller than the number of distinct nodes in the subgraph. If this condition is not fulfilled, a depth-first-search algorithm will search for a loop in the subgraph. If this loop contains shorts only, one short is removed from the subgraph and a new loop is searched for (if any exists). If the loop contains one or more of the other zero-impedance elements, the loop is not allowed. The loop check algorithm signals the first unallowed loop that is found in every subgraph (if any) to the user.

- **Cut-set test**

The dual test detects unallowed cut sets of infinite-impedance elements (independent current source, voltage meter, current-controlled and voltage-controlled current source). Indeed, each of these elements fixes the element's branch current. Hence, in a cut set of n such elements, the n branch currents are determined. Still, Kirchhoff's current law has to be satisfied. As a result, the circuit can be inconsistent, undetermined, redundant or meaningless.

The following, very fast algorithm reports unallowed cut sets. All infinite-impedance elements are removed. The resulting circuit elements are replaced by an undirected branch connecting their n^+ and n^- node. The resulting undirected graph is decomposed into maximally connected subgraphs. Then the following procedure is repeated for every infinite-impedance element. One determines the subgraphs Γ^+ and Γ^- to which the n^+ node and the n^- node of the element belong. If both nodes are a member of different subgraphs, i.e. if $\Gamma^+ \neq \Gamma^-$, then an unallowed cut set is detected and an error is signaled to the user.

After the read-in and the expansion of the circuit, all circuit information is stored in the internal database of the program. If the circuit then successfully passes the topology check, the user is queried to select the analysis type, the circuit excitation, inputs and outputs, as shown in the flowchart of Fig. 3.10. This analysis selection is discussed next.

3.4.3.2. Analysis selection

Within ISAAC, a complete separation has been established between circuit description and analysis control. The circuit topology is described in the textual input file. The user interactively selects the analysis type and assigns the excitation source, the input(s) and output(s) while running the program [GIE_89c].

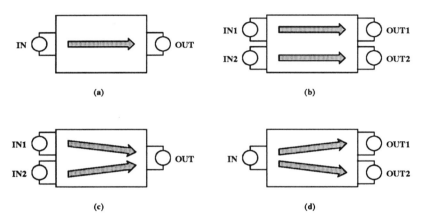

Fig. 3.12. Overview of network functions analysable in ISAAC: (a) transfer function, (b) transfer function ratio, (c) input ratio, and (d) output ratio.

The functions, which can be analyzed directly in ISAAC, not only include conventional transfer functions, but more generally any type of linear network function that is expressed as a ratio of two polynomials. An overview is given in Fig. 3.12. In (a) a transfer function is shown from the signal input (circuit excitation) to the circuit output. The denominator is the system denominator. An impedance or admittance are special cases of a transfer function. The network functions (b), (c) and (d) are ratios of transfer functions in a multivariable system. They are directly generated, without calculation of the system denominator. The general case of a transfer function ratio is shown in (b): both the numerator and denominator inputs and outputs are different. If the numerator and denominator outputs are equal, the general ratio (b) reduces to an input ratio, as shown in (c). Practical examples are the CMRR, the PSRR and the input-referred noise. If the numerator and denominator inputs are equal, the general ratio (b) reduces to an output ratio, as shown in (d). Practical examples are the gain of the second stage of a two-stage amplifier (while the amplifier is excited with a differential input voltage at the first stage) and the loop gain of a feedback structure.

Besides these general options, some frequently used analyzes for opamps are provided: differential-mode gain, common-mode gain, CMRR, power-supply gain, PSRR, and noise analysis. ISAAC automatically recognizes the power supply nodes V_{DD} and V_{SS}. Noise sources are automatically added to resistors and semiconductor devices. Finally, distortion analysis is provided for weakly nonlinear circuits: distortion transfer functions and harmonic distortion ratios [WAM_90]. The noise and distortion analyzes are discussed in chapter 5.

As input (excitation) any voltage or current source can be selected, but also linear combinations of these sources. As output, any node voltage, current through an element, voltage meter or current meter can be asked, but also linear combinations of these. Among other things, this allows the direct simulation of common-mode and differential-mode inputs and outputs. All inputs and outputs can be assigned and reassigned interactively. For example, all of A_{dd}, A_{cd}, A_{cc} and A_{dc} for a fully differential amplifier can be analyzed successively in one interactive simulation run, without editing the circuit description again. For the z-domain analysis of time-discrete circuits, the user also has to specify during which phase the input signal is active and during which phase the output is read off.

The features enumerated above give ISAAC a very broad functionality and enable the simulation of a large number of analyzes on a broad class of analog circuits. After the selection of the analysis type, the circuit excitation, the input(s) and the output(s), three network transformations can be performed, as shown in the flowchart of Fig. 3.10. They will be discussed next.

3.4.3.3. Network transformations

Once the analysis has been chosen, three network transformations can be performed [GIE_89c]. The user interactively selects for each analysis 1) whether mismatch terms are included or not, 2) whether parallel elements are lumped or not, and 3) whether a fully symbolic or a mixed symbolic-numerical analysis is performed. These features greatly improve the interpretability of the result and enable the user to reduce the analysis time as much as possible, since for instance mismatch terms are of no importance for a differential-mode gain, but are for a PSRR. The influence of these transformations on the CPU time and storage is examined more thoroughly in the next chapter. The user can interactively select all these options for each element separately, or for the whole circuit at once. The three transformations are now discussed briefly.

A. Mismatch terms

A first transformation is possible for matching elements, such as transistors M1A and M1B in the CMOS Miller-compensated two-stage opamp of Fig. 3.13. Note that ISAAC recognizes matching elements by their name: matching elements have the same name with a different letter extension, such as M1A and M1B, or R5A and R5B. This matching information is also passed to the components of nonprimitive matching elements.

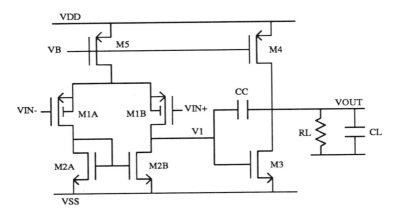

Fig. 3.13. CMOS two-stage Miller-compensated operational amplifier.

In general, there are several possibilities to represent the small-signal components of these matching devices [WAL_89, GIE_89c]. A first possibility is to represent them by different symbols :

$$
\begin{array}{lll}
\text{GM.M1A} \rightarrow \text{GM.M1A} & \text{GM.M1B} \rightarrow \text{GM.M1B} & \\
\text{CGS.M1A} \rightarrow \text{CGS.M1A} & \text{CGS.M1B} \rightarrow \text{CGS.M1B} & (3.16) \\
\dots
\end{array}
$$

This, however, does not take into account that the matching devices are nominally identical. As a result, too many terms are generated which nominally cancel (unless this task is performed by a postprocessor). This also complicates the interpretation of the symbolic expressions and also complicates the heuristic approximation.

Instead of postprocessing a too long expression, the matching information is exploited directly in ISAAC by representing matching elements by the same symbols :

```
GM.M1A → GM.M1          GM.M1B → GM.M1
CGS.M1A → CGS.M1        CGS.M1B → CGS.M1                        (3.17)
...
```

This accelerates the network function calculation and greatly improves the readability of the result.

However, for second-order effects, such as the PSRR, unrealistic results are obtained with (3.17) and device mismatching becomes important. For this, the user can select to include explicit mismatch terms (Δ-terms) in the ISAAC analyzes :

```
GM.M1A → GM.M1          GM.M1B → GM.M1 + ΔGM.M1BA
CGS.M1A → CGS.M1        CGS.M1B → CGS.M1 + ΔCGS.M1BA
...
                                                               (3.18)
```

These mismatch terms are much smaller than the nominal terms. They only slightly increase the analysis time, as will be shown in the next chapter. As compared to the first possibility (3.16), the explicit representation of mismatch terms (3.18) provides the user with more insight into the circuit behavior and yields more realistic error estimations during the approximation of the expressions. This is now illustrated in the following example.

Example
Consider for example the DC value of the PSRR expression for the positive supply voltage for the CMOS Miller-compensated opamp of Fig. 3.13. An approximation with a 25% error yields :

```
                2 GM.M1 GM.M1 GM.M2 GM.M3
─────────────────────────────────────────────────────────────────    (3.19)
  2 GM.M1 GM.M2 GO.M1 GO.M4 + 2 GM.M1 GM.M2 GO.M2 GO.M4
+ ΔGM.M2BA GM.M1 GM.M3 GO.M5 - GM.M1 GM.M3 GO.M1 GO.M5
- ΔGM.M1BA GM.M2 GM.M3 GO.M5 - GM.M1 GM.M3 GO.M2 GO.M5
```

In the denominator, a partial cancellation of two terms, (- GM.M1 GM.M3 GO.M5 (GO.M1 + GO.M2)) and (2 GM.M1 GM.M2 GO.M4 (GO.M1 + GO.M2)), is noticed. These two terms correspond to the contribution of the first stage and the second stage, respectively. Due to this cancellation, a very good *PSRR*+ can be obtained under nominal conditions. However, the mismatch terms (ΔGM.M2BA GM.M1 GM.M3 GO.M5) and (- ΔGM.M1BA GM.M2 GM.M3 GO.M5) also give a contribution, which is of the same order of magnitude as the other terms. This contribution may not be neglected and limits the *PSRR*+ in practice.   ♦

B. Lumping

As a second transformation, like parallel elements can be lumped into one equivalent element and processed as one symbol from then on [GIE_89c, WAL_89]. For example, the parallel capacitors CL and CDB.M1 in the nMOS single-transistor amplifier of Fig. 3.7 can be replaced by one equivalent capacitor CEQ-1 = CL + CDB.M1. Lumped resistors, capacitors and inductors are called GEQ-i, CEQ-i and LEQ-i, respectively, with i being a counting integer. This technique emulates the designer concept of node impedances. Technically, this transformation further reduces computer time and storage, as will be shown in the next chapter.

Example

The gain of the nMOS single-transistor amplifier of Fig. 3.7 with element lumping is given by :

$$
\frac{- \text{ GS } (\text{GM.M1 } - \text{ S CGD.M1})}{\begin{array}{l} \text{GS GEQ-1} \\ + \text{ S (GM.M1 CGD.M1 + GEQ-1 CGS.M1 + GEQ-1 CGD.M1} \\ \quad + \text{ GS CEQ-1 + GS CGD.M1)} \\ + \text{ S}^2 \text{ (CEQ-1 CGD.M1 + CGS.M1 CGD.M1 + CEQ-1 CGS.M1)} \end{array}}
\tag{3.20}
$$

```
        with the following lumped elements being used :
          GEQ-1 = GL + GO.M1
          CEQ-1 = CL + CDB.M1
```

As compared to the formula without lumping (3.13), both the formula complexity and the calculation CPU time have been reduced.   ◆

However, it should be noticed that which elements can be lumped depends on the analysis. For example, if the positive power supply voltage is selected as the input source of the nMOS single-transistor amplifier, GL and GO.M1 are not in parallel anymore and may not be lumped. This also explains why lumping can be different for the numerator and the denominator and why these circuit transformations are introduced after the analysis selection.

C. Symbolic-numerical

A third option provided in ISAAC is to select between a fully symbolic analysis and a mixed symbolic-numerical analysis [GIE_89c]. In the latter case, all elements for which a numerical value is known, are represented by this value. The other elements are represented by a symbol. This yields network functions of type 2. If all elements have been assigned numerical values, the program returns expressions in symbolic form with respect to the complex frequency s or z only (network functions of type 3). If no elements have been assigned numerical values, the program returns fully symbolic expressions (network functions of type 1).

Example
If all elements are represented by their value for a typical design of the CMOS Miller-compensated opamp of Fig. 3.13 (the multi-purpose opamp presented in appendix A with a gain-bandwidth of 1 MHz), the exact expression in s for the $PSRR^+$ becomes :

$$\frac{283587.56 + s\ 10947.523 + s^2\ 20.855215 + s^3\ (-\ 0.373244) + s^4\ 0.000324661}{2.19817 + s\ 127.54224 + s^2\ 20.21014 + s^3\ 1.3461461 + s^4\ 0.015403658}$$

(3.21)

This is a network function of type 3 which completely models the $PSRR^+$ behavior of this operational amplifier, once all device sizes have been determined.   ◆

After carrying out the above circuit transformations, the requested network functions are calculated, approximated and returned to the user, as shown in the ISAAC flowchart of Fig. 3.10. These steps will be described next.

3.4.3.4. Formula calculation and postprocessing
After carrying out the network transformations described above, the linear circuit equations are set up. The desired network functions are then calculated by a dedicated sparse equation solution routine, which will be described in detail in the next chapter. This solution routine returns exact expressions in the complex frequency s or z. These expressions can be displayed or stored in an output file. Usually, these expressions are lengthy and difficult to interpret, due to the large number of symbolic terms. However, taking into account the (order of) magnitude of all symbolic quantities, most of the terms have only a minor contribution to the final result. Therefore, a heuristic approximation is provided, which simplifies the expressions up to a user-defined error percentage [GIE_89c]. This has already been illustrated in the example of the bootstraped bipolar cascode stage in section 3.3.1. The approximation strongly reduces the formula complexity, and retains the dominant contributions only. More details of this approximation algorithm will be presented in the next chapter.

The magnitude of a symbol itself can be determined in four ways [GIE_89c]. First, if the input file contains numerical values, these values are used. Secondly, the magnitude values can be read in from a magnitude file, which contains the DC operating point information obtained by a SPICE [VLA_80b] analysis of the same circuit. Thirdly, ISAAC employs default values for semiconductor small-signal components. These built-in defaults are average values, taking into account the relative magnitude of the different quantities. For example, taking the output conductance of a MOS transistor equal to 1.10^{-6}, the transconductance is set to 100.10^{-6}, which is a reasonable ratio if the transistor operates in the saturation region. Finally, if no value is known, ISAAC will ask the user to supply a value. Note also that the user can supply a value for a whole class of elements at once, since all elements are grouped in element classes (for instance capacitors or voltage-dependent voltage sources). If a particular element has no magnitude value but the element's class has, then this value is inherited by the element. The user can always change any magnitude during the interactive use of the program.

Notice that the selection of an appropriate set of magnitude values is similar to the chicken-and-egg problem. The results of the approximation of course depend on the magnitude values used. If the exact values are unknown (for instance because the circuit still has to be designed, or because a new circuit schematic is entered), default or estimated values have to be used which only reflect the relative sizes of all parameters over a broad range of parameter values. The circuit is then designed based on formulas, which have been simplified for this set of default/estimated parameters. After the design, the parameter values can differ from the default/estimated values. In extreme cases, an approximation with the new set of magnitude values can result in different expressions, and the circuit possibly has to be redesigned. In this context, there are **two ways to use the ISAAC program** :
- if the circuit has already been designed, ISAAC is especially useful for analyzing second-order characteristics (such as the PSRR or the harmonic distortion) with the exact magnitude values (as taken from the DC operating point information supplied by SPICE [VLA_80b]), or for checking the first-order expressions used during the design phase.
- if the circuit has not yet been designed, default or estimated values have to be used to obtain a first-order analytic model which can then be used to size and optimize the circuit [GIE_90b]. A good set of values can always be obtained from a previous design of the circuit for a different but similar set of specifications (if available). Except for really critical designs, this may still result in the right symbolic expressions and thus in good design solutions.

After the approximation, the symbolic expressions are factorized and returned to the user or stored in an output file. After each analysis, the user can simplify the same expressions for another error percentage, as shown in the flowchart of Fig. 3.10, he can ask for another analysis on the same circuit, or he can read in a new circuit.

It is also possible to numerically evaluate the exact or simplified expression over a user supplied frequency range [GIE_89c]. An ASCII table (or an SDIF file) is then generated. The numerical values used for this evaluation are the same as the magnitudes for the approximation. The file can then be sent to a plot program to visualize the circuit behavior. These plots can be compared to SPICE [VLA_80b] output, to obtain an idea of the effective overall error introduced by the approximation.

Example

Fig. 3.14 compares the exact symbolic expression and the 25%-error approximated expression with SPICE simulations of the $PSRR^+$ for a typical design of the CMOS Miller-compensated opamp (as described in appendix A). In these analyzes, the DC operating point information of SPICE has been used and mismatches of 1% have been introduced. Notice how close the ISAAC curves follow the SPICE result up to frequencies far beyond the gain-bandwidth of 1 MHz. Notice also that for the expression with a maximum error percentage of 25%, the effective error for the magnitude of the $PSRR^+$ is much smaller over the whole frequency range. This is due to the error definition used, as will be explained in the next chapter.   ◆

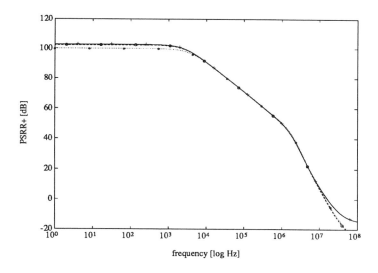

Fig. 3.14. Comparison of SPICE and ISAAC (exact and approximated) results for the $PSRR^+$ of the CMOS Miller-compensated opamp: SPICE result (straight line and '+'), ISAAC 0%-error expression (dashed line and 'o') and ISAAC 25%-error expression (dotted line and '*').

3.5. Conclusions

In this chapter, the symbolic analysis of analog integrated circuits has been discussed in general. Symbolic, numerical and qualitative simulation are three different analysis methods, which give the designer a different perspective on and provide him with complementary information about the same circuit. A symbolic simulator generates the circuit characteristics as analytic expressions in the symbolic circuit parameters, which is, up till now, only feasible for linear (or weakly nonlinear) circuits in the frequency domain. One such a symbolic simulator, ISAAC [GIE_89c], is taken as example and is discussed in more detail. ISAAC is dedicated towards the symbolic analysis of analog integrated circuits. It combines a very broad functionality with a high efficiency, and additionally offers a heuristic approximation of the symbolic expressions.

The main applications of symbolic simulation in analog circuit design have then been discussed. As compared to numerical methods, symbolic analysis provides the designer with more insight into the behavior of the circuit. In combination with a versatile plot program, it also forms a valuable environment for the exploration of existing and new circuit topologies. Another important application is the automatic generation of analytic models used for automatic circuit dimensioning in an automated analog design system. This greatly eases the introduction of new circuit topologies into the system. Finally, symbolic simulation can also be used for the generation of circuit models for behavioral (high-level) simulation, and for the efficient repetitive evaluation of a circuit characteristic (for example in Monte Carlo analysis).

Finally, to illustrate the functionality of present-day symbolic simulators, a general description of the structure and capabilities of the ISAAC program has been given [GIE_89c]. The program analyzes both time-continuous and time-discrete (switched-capacitor) circuits. It accepts all common primitives from circuit theory, and allows the introduction of user-defined subcircuits. Also, the program's flowchart has been discussed. The main modules are the circuit read-in with the expansion and the topology check, the analysis selection, the network transformations (mismatches, lumping and mixed symbolic-numerical), the equation set-up and the equation solution, and the postprocessing with the expression approximation. This has been illustrated with several examples. The detailed algorithmic aspects of the program will now be discussed in the next chapter.

4 ALGORITHMIC ASPECTS OF LINEAR SYMBOLIC SIMULATION

4.1. Introduction

In the previous chapter, the application areas for symbolic simulation have been pointed out and a general description of the structure and functionality of one particular simulator, called ISAAC [GIE_89c, SAN_89], has been given. In this chapter, the algorithmic aspects of symbolic simulation are discussed and illustrated by means of ISAAC: the actual calculation and approximation of the symbolic network functions. In this introductory section, first different network function formats are defined. One of these formats, the canonic sum-of-products format, will be of great importance later on in this chapter. Section 4.2 then gives an historical overview of symbolic analysis and shortly describes the different symbolic analysis techniques presented in the literature. These methods are compared with respect to their usefulness and efficiency. The requirements imposed on the ISAAC program have resulted in the adoption of a determinant calculation method. The set-up of the circuit equations in ISAAC is then discussed in section 4.3. A compacted modified nodal analysis (CMNA) formulation is used, which yields compact and still sparse matrices. The solution of these equations by an efficient sparse determinant expansion algorithm is treated in section 4.4. For semiconductor circuits, however, the exact expressions are usually lengthy and complicated. Section 4.5 describes a heuristic approximation algorithm, which simplifies the symbolic expressions based on the relative magnitudes of the elements and which returns the dominant terms in the result. Section 4.6 then evaluates the performance of the program and compares it to other realizations published in the literature. This comparison will show that the ISAAC program combines a high efficiency with a large functionality for analog circuit designers. Concluding remarks are then provided in section 4.7.

Definition of network function formats

In general, the symbolic network functions for lumped, linear, time-invariant circuits are **rational functions** (ratio of two polynomials) in the complex frequency variable x (s or z) and in the symbolic circuit elements p_j [ADB_80]. Several formats are now defined for these network functions.

Both the numerator and denominator polynomials can be **nested or expanded with respect to the complex frequency and/or the symbolic circuit elements**. A polynomial is called nested when it contains a product of two subpolynomials. In expanded format with respect to the complex frequency, the network functions are given by :

$$H(x) = \frac{N(x)}{D(x)} = \frac{\sum_i x^i . a_i(p_1, \ldots, p_m)}{\sum_i x^i . b_i(p_1, \ldots, p_m)} \qquad (4.1)$$

where the coefficients $a_i(..)$ and $b_i(..)$ are nested or expanded symbolic polynomial functions in the symbolic circuit elements p_j. In fully expanded format, the polynomial coefficients $a_i(..)$ and $b_i(..)$ are in **sum-of-product (SOP) format** and are given by :

$$a_i | b_i = \sum_{k=1}^{r} c_{ik} \left(\prod_{j=1}^{t} p_{ikj} \right) \qquad (4.2)$$

where c_{ik} is a real number and the p_{ikj} are circuit element symbols. A **canonic SOP format** is a SOP format in which each product of symbols appears only once. Hence, no algebraic simplification and no cancellation of terms is possible anymore in a canonic SOP form. A (nested or expanded) polynomial is called **cancellation-free** if its expansion yields a canonic SOP form.

ISAAC internally handles both nested expressions and canonic SOP forms. For an efficient formula approximation, however, canonic SOP forms are required, as will be explained in section 4.5. The results of this approximation are then factorized by the output processor, which can again yield nested expressions. All these definitions are now illustrated in the following examples.

Example 1
The following example polynomials clearly show the difference between nested and expanded expressions :

	nested in s	expanded in s
nested in symbols	$((A + B) C + D S)$ $* (E + F S)$	$(A + B) C E +$ $((A + B) C F + D E) S$ $+ D F s^2$
expanded in symbols	$(A C + B C + D S)$ $* (E + F S)$	$A C E + B C E +$ $S (D E + F A C + F B C)$ $+ s^2 D F$

Example 2

To illustrate the definitions of SOP, canonic SOP, nested and cancellation-free forms, consider the resistive p-circuit of Fig. 4.1. The following four expressions all represent the determinant of the circuit's MNA matrix :

$$(G1 + G2) \ (G2 + G3) \ - \ G2 \ G2 \tag{4.3a}$$

$$G1 \ G2 + G1 \ G3 + G2 \ G2 + G2 \ G3 - G2 \ G2 \tag{4.3b}$$

$$G1 \ (G2 + G3) + G2 \ G3 \tag{4.3c}$$

$$G1 \ G2 + G1 \ G3 + G2 \ G3 \tag{4.3d}$$

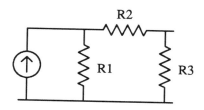

Fig. 4.1. Resistive π-circuit.

Expressions (4.3a) and (4.3c) are nested, whereas expressions (4.3b) and (4.3d) are in SOP form. The SOP forms (4.3b) and (4.3d) are the expansion of the nested forms (4.3a) and (4.3c), respectively. However, both the nested form (4.3a) and the SOP form (4.3b) are not cancellation-free, since the terms (+ G2 G2) and (- G2 G2) cancel in (4.3b). On the other hand, the nested form (4.3c) and the SOP form (4.3d) are cancellation-free. Hence, expression (4.3d) is the canonic SOP form.   ◆

4.2. Overview of symbolic analysis techniques

After these definitions of network function formats, this section discusses symbolic analysis techniques and the development of a symbolic analysis program in general. First, an historical overview of the evolution of symbolic analysis is given. Then, a short overview is presented of the different symbolic analysis techniques found in the literature. They are compared with respect to general applicability and efficiency. Some techniques inherently show many term cancellations, thereby increasing the CPU time and storage requirements. The requirements imposed on the ISAAC program have then resulted in the use of a determinant calculation method.

In the past, many techniques and programs have been presented for the symbolic analysis of linear, lumped, time-invariant circuits [ADB_80, ALD_73, BON_80,

BRU_83, CEN_80,81, CHE_87, CHO_85, CHU_75, COC_73, DEL_89, DOW_69,70ab, FID_73, FRI_74, GEN_74, GIE_89bc,90d, GRI_76, GRU_80, HAR_84, HAS_89a, HUE_89, JOH_84, KON_80,81,88,89, LEE_74,80, LI_85, LIB_88,89, LIN_73ab, MAS_53, MAT_84, MIE_78, MOS_84,86,89, MOU_83, OZG_73, PIE_84, PLI_89, PUN_85, RAA_88, RIE_72, SAN_80,89, SAS_82, SED_88, SHI_74, SIN_74,77, SIU_85, SMI_76,78ab,79,81ab,82, STA_80ab,84,86,89, TAL_65,66, TAN_82, TSA_77, WAL_89,91, WAN_77, XIA_85, YEU_84, ZHA_83ab]. In the sixties and early seventies, symbolic analysis was a major topic in circuit theory and graph theory research. At that time, topological (graph-based) methods were favored above determinant methods [LIN_73a, CHU_75]. Around the mid-seventies, symbolic analysis was widely judged impractical because of the circuit size limitations and CPU time requirements of the earliest programs [CHU_75] and because of the rising success of numerical simulators, such as CANCER [NAG_71] which later on evolved to the well-known SPICE program [VLA_80a]. Since then, however, several techniques have been developed to overcome or reduce these problems, and symbolic analysis has gained a renewed interest in the second half of the eighties. This is partly facilitated by the vastly increasing computing power of the present computers [WAH_89], and is mainly stimulated by the new application domains of computer-aided design and artificial intelligence [KON_89]. At the same time, the superiority of topological methods has decayed. It will be shown experimentally in this chapter that **matrix-based or determinant-based methods can be as good as topological ones (and vice versa).**

Inherent to symbolic analysis and hence to all symbolic analysis techniques, two main problems arise with respect to efficiency. The first problem is the computing time and/or memory storage, which increase very rapidly with the size of the network [CHU_75]. This is mainly due to the **exponential rise of the number of terms with the complexity n of the circuit** ($O(a^n)$ and $O(n^n)$ for some sparse and some dense circuit, respectively, as will be shown in section 4.4.3.1 below). This large number of terms comes from the large number of valid combinations of symbolic elements connected to the $O(n)$ nodes in the circuit. If all these terms are to be generated in an expanded form, the required CPU time increases at best linearly with the number of terms, and thus exponentially with the circuit size. Note that this is true for any symbolic analysis technique, either topological or matrix-based. The problem is even worse for the analysis of linearized circuits. Replacing every transistor by its small-signal model causes a real symbol explosion: the number of branches strongly increases for approximately the same number of nodes (some internal nodes are added). Nevertheless, expanded formulas are still necessary up till now in the presence of term cancellations and for an efficient and reliable approximation, as will be shown in section 4.5 below. Efficient algorithms exist nowadays for the generation and approximation of expanded symbolic network functions. These algorithms are essential for any symbolic analysis program and will be described in this chapter for the ISAAC program [GIE_89c]. In addition, in recent years, several techniques have been developed to further reduce the computer time and memory for large circuits. These techniques include hierarchical decomposition [STA_80b, STA_86], nested expressions [SMI_81b], exploitation of topological knowledge, application of rules, and a clever usage of approximation. They are described in the following chapter. Still, the symbolic analysis of a circuit inevitably takes more time than one numerical simulation of the same circuit.

The second problem is related to **term cancellations**. Two types of cancellations have to be distinguished :
• additive cancellations, in which two equal terms with opposite sign cancel
• multiplicative cancellations, in which a numerator and a denominator factor cancel.
A lot of time and memory is wasted, if intermediate terms/factors are generated which cancel in the final result. These common terms/factors cause an intermediate growth of the number of terms, and also require extra symbolic processing [SMI_76]. As a result, algorithms have to be developed which are inherently cancellation-free [SMI_81b] or techniques have to be applied which detect (as many as possible) invalid symbol combinations before their coefficients are calculated, such as in [ALD_73, MIE_78, SAN_80, SIN_77]. An algorithm is called **inherently cancellation-free** if it generates no term cancellations during the solution of a set of equations with general (i.e. all different, nonzero) symbolic coefficients.

In the literature, **many symbolic analysis techniques have been presented for lumped, linear, time-invariant circuits**. These methods can be classified as follows :
1) matrix manipulation and determinant calculation [GIE_89c, SED_88, SMI_81b, TSA_77]
2) signal-flow-graph method [MIE_78, STA_86]
3) tree-enumeration method [CHO_85, SHI_74]
4) parameter-extraction method [ALD_73, SAN_80]
5) interpolation method [FID_73, SIN_77]
These methods are now described and compared briefly. The task for all these methods is to solve a (general) set of simultaneous linear equations with symbolic coefficients :

$$A.\underline{x} = \underline{b} \tag{4.4}$$

In the case of linear, electronic circuits, this set consists of the circuit equations which describe the linear circuit [SIN_86]. An overview and comparison of the different circuit equation formulations will be presented in the next section. In this section, the different techniques to solve such a set of symbolic equations are described briefly, and are compared with respect to general applicability and efficiency. First, three methods are presented which are matrix-based or determinant-based.

4.2.1. Matrix-manipulation and determinant-calculation methods

These methods generate the network functions by operating on the matrix of the symbolic network equations. If one is interested in all variables $\underline{x}$, then the whole set of equations has to be solved and elimination methods might be appropriate. If one is interested in one variable x_i only, then the application of Cramer's rule is appropriate. This rule implies the calculation of two determinants. Again, there are several algorithms to calculate symbolic determinants. The most important ones are described below.

a) elimination algorithms

By successive transformations, the system of equations $A . \underline{x} = \underline{b}$ is transformed into an equivalent system $A^* . \underline{x} = \underline{b}^*$ in which A^* is an upper triangular matrix (all entries below the diagonal are zero). This equivalent system can then easily be solved by back substitution: first x_n is solved from the last equation, then x_{n-1} is solved by substituting x_n in the forelast equation, etc. These algorithms (such as Gauss and Crout) are the most efficient for the solution of sets of linear equations with numerical coefficients [KUN_86a] and can be used for symbolic equations as well.

The same transformations can also be used to calculate the value of a determinant, since the determinant of a triangular matrix is equal to the product of all matrix entries on the diagonal.

Example

The determinant of the following 3x3-matrix M with general entries is calculated with the elimination algorithm :

$$M = \begin{bmatrix} A & B & C \\ D & E & F \\ G & H & I \end{bmatrix} \hspace{3cm} (4.5)$$

The entries in the first column under the diagonal can be made zero by multiplying the second (third) row with A and subtracting D (G) times the first row. Since both the second and the third row have been multiplied with A, the result must be divided by A^2 to leave the value of the determinant unchanged. This yields :

$$|M| = \frac{1}{A^2} \begin{vmatrix} A & B & C \\ 0 & \begin{vmatrix} A & B \\ D & E \end{vmatrix} & \begin{vmatrix} A & C \\ D & F \end{vmatrix} \\ 0 & \begin{vmatrix} A & B \\ G & H \end{vmatrix} & \begin{vmatrix} A & C \\ G & I \end{vmatrix} \end{vmatrix}$$ (4.6)

Applying the same technique to make the entry in the second column under the diagonal zero yields :

$$|M| = \frac{1}{A^2 \begin{vmatrix} A & B \\ D & E \end{vmatrix}} \begin{vmatrix} A & B & C \\ 0 & \begin{vmatrix} A & B \\ D & E \end{vmatrix} & \begin{vmatrix} A & C \\ D & F \end{vmatrix} \\ 0 & 0 & \begin{vmatrix} \begin{vmatrix} A & B \\ D & E \end{vmatrix} & \begin{vmatrix} A & C \\ D & F \end{vmatrix} \\ \begin{vmatrix} A & B \\ G & H \end{vmatrix} & \begin{vmatrix} A & C \\ G & I \end{vmatrix} \end{vmatrix} \end{vmatrix}$$ (4.7)

Hence, the resulting determinant is given by :

$$|M| = \frac{1}{A^2 \begin{vmatrix} A & B \\ D & E \end{vmatrix}} \star A \begin{vmatrix} A & B \\ D & E \end{vmatrix} \begin{vmatrix} \begin{vmatrix} A & B \\ D & E \end{vmatrix} & \begin{vmatrix} A & C \\ D & F \end{vmatrix} \\ \begin{vmatrix} A & B \\ G & H \end{vmatrix} & \begin{vmatrix} A & C \\ G & I \end{vmatrix} \end{vmatrix}$$ (4.8)

The common factors cancel between the numerator and the denominator. Elaborating the result yields :

$$|M| = \frac{1}{A} [(A E - B D)(A I - G C) - (A H - B G)(A F - D C)]$$ (4.9)

The denominator factor A cannot be canceled. However, the common terms +BDGC and -BGDC cancel and now factor A does cancel. The final result is :

$$|M| = A E I - E G C - B D I - A H F + H D C + B G F$$ (4.10)

The same result could be obtained directly with a determinant expansion method. However, to obtain the cancellation-free form (4.10) with the elimination method, many multiplicative and additive cancellations have to be performed. This implies a lot of symbol processing on and an expansion of the intermediate results, thereby strongly increasing the CPU time above the limit $O(n^3)$ for numerical determinants. The elimination method obviously is not cancellation-free and hence is as such less efficient for the solution of symbolic equation sets.   ◆

An improvement to the simple elimination algorithm which generates no fractions during the subsequent elimination steps is called the fraction-free method. A further improvement which suppresses the indermediate expression expansion also for determinants with multivariate polynomial entries is presented in [SAS_82]. However, the authors admit that their method is only comparable to the minor expansion method for the calculation of dense multivariate determinants and particularly for the solution of sets of dense symbolic linear equations. For sparse determinants or sparse equation sets, which typically is the case for linear circuits, the minor expansion method is superior [SAS_82]. This method is discussed next.

b) recursive determinant-expansion algorithms

A determinant can also be calculated by means of a recursive expansion formula [GRI_76, SMI_79, WAN_77]. These algorithms are free of additive and multiplicative cancellations for general matrix entries (i.e. if all matrix entries are different) [SMI_81b].

The Laplace expansion of the determinant of the nxn-matrix A along an arbitrary row i is given by :

$$|A| = \sum_{j=1}^{n} a_{ij} (-1)^{i+j} |M_{ij}| \qquad \text{for } (n > 1) \qquad (4.11)$$

$$= 1 \qquad \text{for } (n = 1)$$

A similar expression exists for development along an arbitrary column. The minors $|M_{ij}|$ are the determinant of the $(n-1)x(n-1)$-matrix obtained from the original matrix A by removing row i and column j. These minors are recursively calculated with the same formulas.

Note that in the most general case the determinant can recursively be expanded along multiple rows or columns at once. It will be shown, however, in section 4.4 below that in general this has no speed improvement above the development along a single row or column.

The recursive expansion method (4.11) can be improved further and turned into an efficient algorithm for the evaluation of sparse, symbolic determinants, as will be described further on in section 4.4.

Example

The determinant of the following *3x3*-matrix M with general entries is developed along the first row, resulting in :

$$|M| = \begin{vmatrix} A & B & C \\ D & E & F \\ G & H & I \end{vmatrix}$$

$$= A \begin{vmatrix} E & F \\ H & I \end{vmatrix} - B \begin{vmatrix} D & F \\ G & I \end{vmatrix} + C \begin{vmatrix} D & E \\ G & H \end{vmatrix}$$

$$= A\,E\,I - A\,H\,F - B\,D\,I + B\,G\,F + C\,D\,H - C\,G\,E \qquad (4.12)$$

This is the same result as with the elimination method (4.10). However, the cancellation-free form is now obtained directly, without any intermediate cancelling terms or products. This can easily be explained because at each recursion level in the determinant expansion, an element is multiplied with a minor which does not contain this element anymore neither any of the other elements of the same column (in the case of column expansion). As a result, the determinant expansion method is inherently cancellation-free. In practice, cancellations can still occur if not all matrix entries are different. These cancellations are then caused by the particular circuit equation formulation used, and not by the equation solution algorithm.   ◆

The recursive determinant-expansion method is now illustrated for a practical example, which will also be solved with other methods later on in this section and allows us to compare these methods.

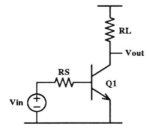

Fig. 4.2. Bipolar single-transistor amplifier.

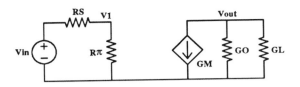

Fig. 4.3. Equivalent small-signal circuit of the bipolar single-transistor amplifier.

Example

Consider the DC gain of the bipolar single-transistor amplifier of Fig. 4.2. The equivalent small-signal circuit is shown in Fig. 4.3. The CMNA circuit equations are given by :

$$\begin{bmatrix} GS + G\pi & 0 \\ GM & GO + GL \end{bmatrix} \begin{bmatrix} V_1 \\ V_{out} \end{bmatrix} = \begin{bmatrix} GS \\ 0 \end{bmatrix} V_{in} \tag{4.13}$$

Solving these equations with Cramer's rule and with the above determinant expansion algorithm immediately yields :

$$\frac{V_{out}}{V_{in}} = \frac{- GM \; GS}{GS \; GL + G\pi \; GL + GS \; GO + G\pi \; GO} \tag{4.14}$$

without any term cancellations for this example.   ◆

c) nonrecursive nested-minors method

The nested-minors method performs a nonrecursive Laplace expansion of the determinant along a column (or row) in a bottom-up way [GEN_74]. First, all possible 2x2-minors from 2 arbitrary columns (rows) are calculated. Then these minors are used to calculate all possible 3x3-minors from these 2 columns (rows) and one additional column (row). These minors are then used to calculate all possible 4x4-minors from the three previous columns (rows) and one additional column (row), and so forth. This expansion by column (row) minors is a special case of the general nonrecursive Laplace expansion along multiple columns (rows) at once. However, it is proven in [GEN_74] that the expansion by column (row) minors has the least cost (in multiplications) from all such nonrecursive minor expansion methods (if all multiplications have an equal cost in CPU time).

Example

The determinant of the following *4x4*-matrix N with general entries is expanded by row minors starting from the 2 bottom rows, yielding :

$$
|N| = \begin{vmatrix} A & B & C & D \\ E & F & G & H \\ I & J & K & L \\ M & N & O & P \end{vmatrix}
$$

$$
\begin{aligned}
= & \; A \; [F \; (K \; P \; - \; O \; L) \; - \; G \; (J \; P \; - \; L \; N) \; + \; H \; (J \; O \; - \; K \; N)] \\
& - B \; [E \; (K \; P \; - \; O \; L) \; - \; G \; (I \; P \; - \; L \; M) \; + \; H \; (I \; O \; - \; K \; M)] \\
& + C \; [E \; (J \; P \; - \; L \; N) \; - \; F \; (I \; P \; - \; L \; M) \; + \; H \; (I \; N \; - \; J \; M)] \\
& - D \; [E \; (J \; O \; - \; K \; N) \; - \; F \; (I \; O \; - \; K \; M) \; + \; G \; (I \; N \; - \; J \; M)]
\end{aligned} \qquad (4.15)
$$

From the 12 *(2x2)*-minors needed, only 6 are different. Indeed, from the 2 bottom rows and the 4 columns, only 6 *(2x2)*-minors can be formed. Hence, by precalculating and reusing these minors, the total determinant calculation time is reduced.

♦

However, it is clear that this technique is not optimal for sparse matrices, since it calculates all *(2x2)*-, *(3x3)*-... minors before it is known if these minors are really needed. This method is useful for dense matrices only [SMI_76].

All methods described above, the eliminination method, the determinant-expansion method and the method of the nested minors, are matrix-based or determinant-based. The following two techniques are graph-based.

4.2.2. Signal-flow-graph method

The circuit equations are now represented by the corresponding signal-flow graph, a weighted, directed graph with the variables as nodes and the circuit elements as branch weights [LIN_73a]. Any transfer function from an input variable x_i to an output variable x_j can then be obtained from the structure of this signal-flow graph by means of **Mason's rule** [MAS_53] :

$$
\frac{x_j}{x_i} = \frac{\Sigma \; P_k \; \Delta_k}{\Delta} \qquad (4.16)
$$

where :

$\Delta = 1$ - (sum of all loop weights) + (sum of all second-order loop weights) - (sum of all third-order loop weights) + ...

P_k = weight of the *k*-th path from the input node x_i to the output node x_j

Δ_k = sum of those terms in Δ without any constituent loops touching P_k

and the summation is taken over all paths from x_i to x_j. An nth-order loop is a set of n nontouching loops. A path (loop) weight is the product of all branch weights along the path (loop).

The evaluation of a network function with this method entails the enumeration of all paths (from x_i to x_j) and all orders of loops in the signal-flow graph. Algorithms for this path and loop enumeration can be found in [CHU_75].

The signal-flow-graph method can easily be shown to be inherently cancellation-free if the circuit equations are written in the following form :

$$\underline{x} = A.\underline{x} + \underline{b} \tag{4.17}$$

It is always possible to formulate the circuit equations in this form, and to construct the corresponding signal-flow graph directly from the circuit description, by selecting a tree, considering the tree branch voltages and the cotree branch currents as variables, and applying the KCL, KVL and constitutive equations [CHU_75]. Note, however, that the total number of all orders of loops and also the number of term cancellations strongly depend on the choice of the tree used to formulate the signal-flow graph [LIN_73a].

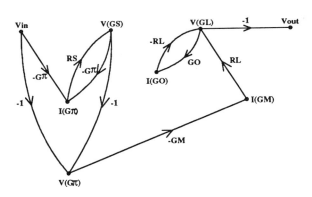

Fig. 4.4. Signal-flow graph for the small-signal circuit of the bipolar single-transistor amplifier.

Example
Consider again the DC gain of the bipolar single-transistor amplifier of Fig. 4.2, with the equivalent small-signal circuit shown in Fig. 4.3. If the tree {vin, GS, GL} is selected, then the corresponding signal-flow graph is depicted in Fig. 4.4. This signal-flow graph contains two nontouching loops :

$$L_1 = - \text{RS G}\pi$$

$$L_2 = - \text{RL GO}$$

(4.18)

Hence, the system determinant Δ is given by :

$$\Delta = 1 - L_1 - L_2 + L_1 L_2$$

$$= 1 + \text{G}\pi \text{ RS} + \text{RL GO} + (\text{RS G}\pi) \ (\text{RL GO})$$

(4.19)

The signal-flow graph also contains two paths from V_{in} to V_{out} :

$$P_1 = (- \text{ G}\pi) \text{ RS} \ (-1) \ (- \text{ GM}) \text{ RL} \ (-1)$$

$$P_2 = (-1) \ (- \text{ GM}) \text{ RL} \ (-1)$$

(4.20)

Since the second path P_2 does not touch loop L_1, the numerator of the transfer function is given by :

$$\Sigma \ P_k \ \Delta_k = P_1 + P_2 \ (1 - L_1)$$

$$= \text{G}\pi \text{ RS GM RL} - \text{GM RL} \ [1 + \text{RS G}\pi]$$

(4.21)

In this expression, two terms cancel. Also, the interpretation of the results is troubled by the mixture of impedances and admittances. Elements which accidentally belong to the tree are represented in impedance form, elements from the cotree in admittance form. After carrying out the cancellations and converting the result to all-admittance form, the following result is obtained :

$$\frac{V_{out}}{V_{in}} = \frac{- \text{ GM GS}}{\text{GS GL} + \text{G}\pi \text{ GL} + \text{GS GO} + \text{G}\pi \text{ GO}}$$

(4.22)

which is the same as in (4.14) with the determinant-expansion method. On the other hand, if the tree $\{v_{in}, \text{ G}\pi, \text{ GL}\}$ is selected, no term cancellations occur in this example. In practice, for larger examples, still many term cancellations occur even for the best tree, and no general method exists for the selection of an optimal tree for each application [LIN_73a]. ◆

An alternative signal-flow-graph-based formula for the generation of symbolic network functions has been presented in [MIE_78]. It produces less term cancellations and is less sensitive to the selected network tree than Mason's formula. On the other hand, it generates still more unvalid symbol combinations than for instance the algebraic method reported in [SAN_80].

Since the signal-flow-graph method derives the network function from the structure of some graph associated with the system, it is called a **topological method**. Another topological method is the tree-enumeration method, to be discussed next.

4.2.3. Tree-enumeration method

The tree-enumeration method is based on a theorem which holds for any square matrix with the property that the sum of all entries in each row and column is zero. Hence, the circuit equations have to be defined by means of the indefinite admittance matrix (IAM) Y_{ind} which shows this zero-sum property [CHU_75] :

$$\underline{i} = Y_{ind} \cdot \underline{v} \tag{4.23}$$

The IAM relates the n terminal currents $\underline{i}$ and the n terminal voltages $\underline{v}$ of an n-terminal network. Note that this method holds for RLC-g_m circuits only, and that no reference node has been selected yet in the IAM. The IAM equations are then represented by a directed, weighted graph, constructed directly from the circuit schematic by replacing every element by a proper set of directed branches [CHU_75].

Any network function is then calculated by means of the following theorem [CHU_75]: for any RLC-g_m network, the determinant of the nodal-admittance matrix or the system determinant (with any node as the reference node) is equal to the sum of the branch-weight products of all directed trees (with any node as the root) in the directed graph of the circuit. Hence, the evaluation of a network function with this method entails the enumeration of all directed trees in the directed graph (after a root has been selected). Algorithms for enumerating trees can be found in [ADB_80].

As already stated above, the tree-enumeration method is not a general method, since it can be applied to RLC-g_m networks only. In addition, this method is inherently not cancellation-free, because of the zero-sum property which implies that each element XYZ appears in exactly four positions in the IAM according to the following pattern :

$$\begin{bmatrix} & \cdots & & & \cdots & \\ \cdots & +XYZ & \cdots & -XYZ & \cdots \\ & \cdots & & & \cdots & \\ \cdots & -XYZ & \cdots & +XYZ & \cdots \\ & \cdots & & & \cdots & \end{bmatrix} \tag{4.24}$$

This results in many term cancellations, implying that directed-tree branch-weight products are generated, which are to be canceled in the final result. Also, the number of trees strongly increases with the size of the network. This limits the tree-enumeration method to about the same circuit size as the signal-flow-graph method

[CHU_75]. Since in addition the tree-enumeration method is not generally applicable and since it inherently suffers from term cancellations, this method is inferior to the signal-flow-graph method.

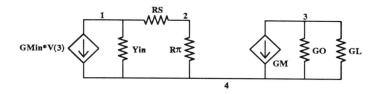

Fig. 4.5. Augmented small-signal circuit of the bipolar single-transistor amplifier for the calculation of the gain.

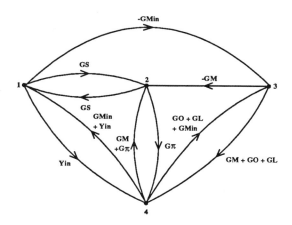

Fig. 4.6. Directed graph for the augmented small-signal circuit of the bipolar single-transistor amplifier.

Example

Consider again the DC gain of the bipolar single-transistor amplifier of Fig. 4.2. To obtain both the numerator and the denominator of the gain in one calculation, the circuit is augmented with the elements GMIN and YIN at the input [CHU_75], as shown in Fig. 4.5. The directed, weighted graph of this augmented circuit is shown in Fig. 4.6. If node 4 is selected as the root, 9 directed trees can be detected in this graph. The sum of the corresponding branch-weight products contains 21 terms, 16 of which cancel. The result is then sorted according to GMIN and YIN to obtain the following transfer function :

$$\frac{V_{out}}{V_{in}} = \frac{- \; GM \; GS}{GS \; GL \; + \; G\pi \; GL \; + \; GS \; GO \; + \; G\pi \; GO} \qquad (4.25)$$

which is the same as in (4.14) and (4.22) with the determinant-expansion method and the signal-flow-graph method, respectively.   ◆

After these graph-based methods, two techniques are described which combine numerical and symbolic techniques to calculate symbolic network functions.

4.2.4. Parameter-extraction method

In the parameter-extraction method, the circuit equations have to be formulated according to the indefinite admittance matrix (IAM) Y_{ind}. The evaluation of any cofactor of this IAM containing the symbolic parameter α is now reduced to the evaluation of the cofactor of two new IAMs not containing α, according to the following theorem [ALD_73]. Let α be a symbol appearing in the *(nxn)*-IAM Y_{ind} in four positions: $+\alpha$ in positions *(i,k)* and *(j,m)* and $-\alpha$ in positions *(i,m)* and *(j,k)*. Then :

cofactor of Y_{ind} = cofactor of $(Y_{ind} \, | \, \alpha=0)$

$$+ \; (-1)^{j+m} \; \alpha \; (\text{cofactor of } Y_{\alpha}) \qquad (4.26)$$

where Y_{α} is an *(n-1)x(n-1)*-IAM not containing α, obtained by modifying Y_{ind} as follows :
• add row *j* to row *i*
• add column *m* to column *k*
• delete row *j* and column *m*

If $(Y_{ind} \, | \, \alpha=0)$ and Y_{α} contain other symbols, the theorem is applied repeatedly to extract all symbols. If a matrix is obtained with numerical entries only, the cofactor is evaluated by any standard numerical method (for example Gaussian elimination). In this way, all combinations of symbols are generated with the appropriate coefficient. In [ALD_73], a method is described to generate the valid symbol combinations only. Still, the number of valid symbol combinations can be very large, and for each of them the corresponding numerical coefficient has to be determined separately by numerically calculating the cofactor of some IAM. Nevertheless, the parameter-extraction method is suited for the analysis of large circuits where only a few elements are represented by symbols (network functions of type 2) [CHU_75].

4.2.5. Polynomial interpolation method

This method is based on the well-known theorem that the univariate polynomial $P(x)$ of degree n is uniquely defined by the values $P(x_i)$ at $n+1$ distinct values x_i and can be reconstructed by interpolating these values. This technique is then extended to the calculation of symbolic, multivariate network functions [FID_73]. The calculation of the values $P(x_i)$ and the interpolation itself are performed with fast numerical algorithms. However, care must be paid to the numerical accuracy and the error propagation of these numerical algorithms, which can result in erroneous interpolation results [SMI_81b].

4.2.6. General comparison of the different techniques

All techniques presented above can be used and have been used for the symbolic analysis of linear, lumped, time-invariant circuits. They are now compared with respect to general applicability and efficiency.

Both the determinant-expansion and signal-flow-graph methods are generally applicable and are inherently cancellation-free (for general matrix entries). In practice, due to the particular matrix fill-in of the circuit equation formulation used, term cancellations can still occur with these methods. The tree-enumeration method, on the other hand, is only applicable to $RLC\text{-}g_m$ circuits and inherently shows many term cancellations. The parameter-extraction method is appropriate if only a few network elements are represented by a symbol. The interpolation method is very sensitive to numerical errors.

Hence, for a general, robust and efficient symbolic analysis program, only the determinant-expansion method and the signal-flow-graph method are appropriate. The signal-flow-graph approach was very popular in the seventies [CHU_75] and can take advantage of all results from graph theory research, but has the disadvantage that the circuit equations have to be set up in a special way (which is actually a limitation), and that the corresponding number of term cancellations depends on the tree selected to formulate these circuit equations. Furthermore, the selection of an optimal tree for a circuit is not straightforward. The only rule here is that the tree should resemble a star tree as much as possible [LIN_73a]. Determinant-based methods, on the other hand, can be applied to any circuit, to any circuit equation matrix formulation and more general to any matrix in a straight way. And although determinant-based methods have for a long time been considered impractical and inferior to topological methods, the research results presented in this chapter will experimentally demonstrate that determinant-based methods, if implemented efficiently, can be as good as topological ones. Besides, there must always be some equivalence relation between a matrix-based and a graph-based formulation and solution method of the same problem, and vice versa.

Which symbolic analysis technique to use for a particular implementation of a symbolic analysis program, depends on all boundary conditions or requirements imposed on the program, and has to be selected to satisfy the final goal of any symbolic analysis program: the generation of symbolic network functions in the most compact form and in the most efficient way for the application domain at hand. From the point of efficiency for example, cancellation-free methods are preferred over noncancellation-free methods. **For the ISAAC program [GIE_89c], this has resulted in the following algorithmic decisions.** The program has been directed towards analog circuit designers. As symbolic analysis strategy, a determinant calculation method has been adopted. The network equations are formulated according to the compacted modified nodal analysis (CMNA) method, which shows a fair compromise between matrix size and additional term cancellations. The determinants are calculated by recursive minor expansion exploiting the sparseness of the equations. The error value for the heuristic approximation can be varied continuously (between 0 and 100%) and the approximation itself can be carried out over the whole frequency range. These two conditions require the expansion of all expressions, although of course a nested form is more compact and can be calculated in a shorter CPU time. All these options are described in more detail in the following sections. They are also compared to alternatives, in order to reveal the appropriateness and efficiency of these choices. In the next section, the set up of the circuit equations is discussed.

4.3. The set-up of the linear circuit equations

The calculation of symbolic network functions in ISAAC is performed in three consecutive steps. First, the linear compacted modified nodal analysis (CMNA) equations are set up, which describe a lumped, linear, time-invariant electrical circuit in the complex frequency domain. The same formalism is used both for time-continuous and time-discrete circuits. The circuit equation set-up is described in this section. Next, the CMNA equations are solved by a dedicated, sparse equation solution algorithm to obtain the desired network function. This algorithm is described and compared to alternatives in the next section 4.4. The network function is then simplified with a heuristic approximation algorithm, presented in section 4.5. Finally, the performance of the program is evaluated and compared to other realizations in section 4.6.

To minimize the CPU time consumption of the equation solution algorithm, a **Compacted Modified Nodal Analysis (CMNA)** method is used for the formulation of the circuit equations. This formalism yields compact and still relatively sparse equations, and is compared to other circuit equation formulations in subsection 4.3.1. The set-up of s-domain CMNA equations for time-continuous circuits is then treated in subsection 4.3.2. The set-up of z-domain CMNA equations for time-discrete circuits is presented in subsection 4.3.3.

4.3.1. Comparison of circuit equation formulations

The method by which the circuit equations are formulated is of key importance to any symbolic analysis program. It strongly effects the execution speed and storage requirements of the equation solution algorithm, both through the matrix size, the matrix sparseness and the term cancellations introduced by the fill-in of the matrix entries (which are characteristic for each particular formulation). In the following, several formulations are compared with respect to their general applicability, the matrix size, the matrix sparseness and the possible introduction of term cancellations. The **sparseness of a matrix** is defined as the ratio of the number of zero entries to the total number of matrix entries.

4.3.1.1. Tableau approach

A first general approach for linear, time-invariant circuits is based on the tableau matrix [CHU_87]. It consists of combining a set of linearly independent KCL equations, a set of linearly independent KVL equations and the branch relations (BR) into one global matrix. All circuit elements are represented by a branch (controlled sources by two branches). An arbitrary node is selected as the reference node. The variables in the equations are all node voltages, all branch voltages and all branch currents.

In matrix form, the linear tableau equations are written as [CHU_87] :

$$
\begin{array}{c} \text{KCL} \\ \text{KVL} \\ \text{BR} \end{array}
\begin{bmatrix} 0 & 0 & A \\ -A^T & 1 & 0 \\ 0 & M & N \end{bmatrix}
\begin{bmatrix} \underline{e} \\ \underline{v} \\ \underline{i} \end{bmatrix}
=
\begin{bmatrix} 0 \\ 0 \\ \underline{u}_s \end{bmatrix}
\qquad (4.27)
$$

where A is the reduced incidence matrix (obtained by applying KCL on every node but the reference node), $\underline{i}$ is the vector of branch currents, $\underline{v}$ the vector of branch voltages and $\underline{e}$ the vector of all node voltages (with respect to the reference node), and $\underline{u}_s$ is the vector of the independent (current and voltage) sources. The branch equations are written in general form in (4.27) with admittance-type elements in M and impedance-type elements in N. If the circuit graph contains n nodes and b branches, then the dimensions of the tableau matrix T are $[(n-1)+2b]x[(n-1)+2b]$.

Example

The active RC filter of Fig. 4.7 consists of 11 nodes and 22 branches. The tableau matrix is a 54x54 matrix with a sparseness of 95.6%. The situation is even worse for semiconductor circuits which have a much larger number of branches for the same number of nodes due to the small-signal expansion. For example, the CMOS OTA of Fig. 4.8 is described by a 95x95 tableau matrix with a sparseness of 97.3%.   ◆

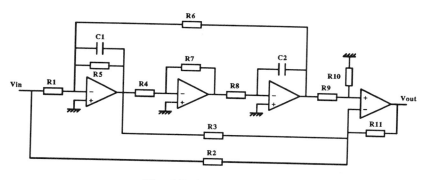

Fig. 4.7. Active RC filter.

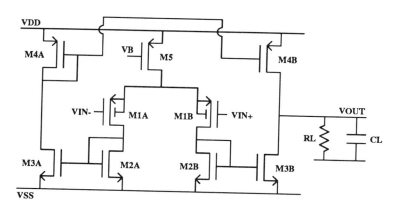

Fig. 4.8. CMOS operational transconductance amplifier (OTA).

It can be concluded that the tableau matrix is large and sparse, even for small circuits. This may cause no problems for numerical analysis if an efficient numerical sparse-matrix algorithm is available. However, the large matrix size imposes a problem for symbolic analysis, also if sparse-matrix datastructures are used. This will be explained in section 4.4 when the equation solution algorithm is discussed. Still, some symbolic analysis programs from the literature use the tableau (or a similar) approach, such as in [SIN_77] or [SMI_81a]. Indeed, the tableau formulation has the advantage that each symbolic circuit parameter appears only in one position in the tableau matrix, namely in the submatrix M or N. The reduced incidence matrix A contains only entries +1, -1 and 0. Hence, each symbol appears only at one position in the tableau matrix, on the condition that all branches are represented by a different symbol (which makes the exploitation of matching impossible). As a result, if a cancellation-free determinant algorithm is applied to the tableau matrix and under the above condition of no matching exploitation, no cancellations occur during the equation solution. On the other hand, much CPU time is lost due to the large matrix size, especially for symbolic determinants. Moreover, the above advantage disappears when matching information is exploited and different elements are represented by the same symbols.

4.3.1.2. Modified nodal analysis (MNA)

A more compact formulation is the nodal analysis (NA) method which considers only the node voltages as variables. The conventional NA method, however, does not allow all basic circuit primitives [CHU_87]. The modified nodal analysis (MNA) method [HO_75] on the other hand combines the compactness of the NA method with the generality of the tableau matrix. It allows all primitives with polynomial matrix entries at the expense of a somewhat larger matrix size. The variables involved are all node voltages (with respect to an arbitrary reference node), and some selected branch currents. In this method, the KCL equations of all nodes are written in terms of the node voltages, but whenever an inductor is encountered or an element that is not voltage-controlled, the corresponding branch current is added as a new variable, and the branch relation of the element is added as a new equation. Hence, currents are introduced for an inductor, an independent voltage source, a voltage-controlled voltage source, a current-controlled current source, a current-controlled voltage source (both the controlled and controlling currents), a current meter, a short, a nullor, a norator, and an opamp.

In matrix notation, the MNA equations are written as [CHU_87] :

$$
\begin{array}{c} \text{KCL} \\ \text{BR} \end{array}
\begin{bmatrix} A & B \\ C & D \end{bmatrix}
\begin{bmatrix} \underline{e} \\ \underline{i} \end{bmatrix}
=
\begin{bmatrix} \underline{J} \\ \underline{E} \end{bmatrix}
\tag{4.28}
$$

where $\underline{e}$ is the vector of the node voltages, $\underline{i}$ the vector of the selected branch currents, $\underline{J}$ the vector of the independent current sources, and $\underline{E}$ the vector of the independent voltage sources. The size of the MNA matrix is $[(n-1)+m]x[(n-1)+m]$, where m is the number of selected branch currents and usually $m << b$ (b being the number of branches). The matrix can easily be filled in directly from the circuit description by means of the element stamps. The major drawback of the MNA method, however, as compared to the tableau method, is that the same symbolic parameter can appear on different positions in the MNA matrix, sometimes with plus sign, sometimes with minus sign, even if all elements are represented by a different symbol. This can result in term cancellations during the equation solution afterwards. The matrix size on the other hand is much smaller than for the tableau description.

Example

The MNA matrix for the active RC filter of Fig. 4.7 is now 15x15. Five currents have to be added. The sparseness is 80.9%. For the CMOS OTA of Fig. 4.8 the size of the MNA matrix is 9x9 with a sparseness of 64.2%.   ◆

As compared to the MNA method, however, the circuit equations can still be formulated in a more compact way, yielding smaller matrix sizes and less term cancellations, while still being generally applicable and maintaining the polynomial character of all matrix entries. This results in the compacted MNA, which is discussed next.

4.3.1.3. Compacted modified nodal analysis (CMNA)

The CMNA method [GIE_89c, WAL_89] is derived from the MNA method. It is the most common-sense method as it only includes the variables and equations which are strictly necessary to calculate the requested network function. In this sense, it strongly resembles the formulations [NAG_71] which were used before the MNA method was generally adopted for numerical simulation (such as in SPICE [VLA_80a]). The CMNA method consists of applying the KCL equations to all nodes (but the reference node) and some additional branch relations. However, not all MNA variables are included in the CMNA equations. By applying some compactions on the MNA equations, some variables (both node voltages and branch currents) and some equations disappear. One ends up with less equations in less variables, some of which are node voltages, some of which are selected branch currents. These compactions are now described briefly.

• *Row compactions*
   A first compaction is applied to the rows of the MNA matrix. In the CMNA method, additional branch currents are introduced only if strictly necessary. So, only the current through a current meter, the controlling current of a current-controlled current and voltage source, and the controlled current of a current-controlled and voltage-controlled current source are added. All other currents (the current through an independent voltage source, a current-controlled and voltage-controlled voltage source, a short or a norator) are of no importance for the transfer function from input to output, unless this current is (part of) the output. In that case the current is added as one extra variable. Also, the current of a selected input voltage source is added as extra variable to enable the direct simulation of the input admittance. Note that these options pose no problems in ISAAC since the selection of the input and the output precede the set-up of the circuit equations and the corresponding circuit matrix (see the ISAAC flowchart in the previous chapter).

Summarizing, several MNA currents are omitted in the CMNA equations without loosing any information about the requested network function. For the primitives involved, only one KCL equation is written for both terminal nodes n_+ and n_-

together (see Fig. 4.9). The resulting equation is the sum of the KCL equations for n_+ and n_- of the original MNA formulation. In these MNA equations, the branch current appears once with a coefficient +1 and once with a coefficient -1. In the sum of both equations, the current disappears. As a result, since some KCL equations are applied to several nodes together, the net number of KCL equations and additional branch currents is smaller than for the MNA method.

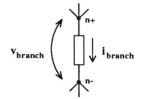

Fig. 4.9. A circuit branch.

- **Column compactions**

A second compaction is applied to the columns of the MNA matrix. In the CMNA method, the number of node voltages is reduced as well without loosing any information. Indeed, both terminal node voltages of a current meter, a short and a nullor (also at the input of an ideal opamp) are equal. Hence, these voltages are two by two represented by one variable only and the corresponding trivial branch relations of the type :

$$v_+ - v_- = 0 \qquad\qquad (4.29)$$

are omitted. Each time v_+ or v_- is addressed in the circuit equations, the same variable and the same column in the CMNA matrix is accessed.

A similar technique is possible for a voltage source. Both terminal node voltages differ only by a constant voltage E. If the voltage source is not excited, both node voltages are equal. Hence, the branch relation of the type :

$$v_+ - v_- = E \qquad\qquad (4.30)$$

is omitted. Both node voltages v_+ and v_- now refer to the same column in the CMNA matrix. If the voltage source is excited, a second column is added, the right-hand side vector corresponding to the excitation E. Summarizing, by the elimination of node voltages in terms of other node voltages, the net number of node voltage variables and branch relations is smaller than for the MNA method.

Both compactions reduce the number of rows and the number of columns. In matrix form, the CMNA equations are written as [GIE_89c] :

$$\begin{array}{c} \text{KCL} \\ \\ \text{BR} \end{array} \left[\begin{array}{cc} A' & B' \\ \\ C' & D' \end{array} \right] \left[\begin{array}{c} \underline{e}' \\ \\ \underline{i}' \end{array} \right] = \left[\begin{array}{c} \underline{J}' \\ \\ \underline{E}' \end{array} \right] \qquad (4.31)$$

where $\underline{e}'$ is the reduced vector of the node voltages, $\underline{i}'$ the vector of the additional branch currents, $\underline{J}'$ the vector of the independent current sources, and $\underline{E}'$ the vector of the independent voltage sources. The structure is the same as of the MNA equations (4.28). The size of the CMNA matrix is $[(n-1-p)+q]x[(n-1-p)+q]$, where p is the number of column compactions and q is the number of additional CMNA branch currents. Due to the compactions this matrix size is much smaller than for the MNA method. However, since the compactions are different for rows and columns, the symmetry between rows and columns is broken and whereas A and D in (4.28) are square matrices, A' and D' in (4.31) in general are not. Again, every symbolic element can appear on several positions. However, as compared to the MNA method, the number of term cancellations is reduced due to the compactions, as will be illustrated by the example in the next subsection.

Example
The CMNA matrix for the active RC filter of Fig. 4.7 is now 5x5. No extra current has to be added. The sparseness is 56.0%. For the CMOS OTA of Fig. 4.8 the size of the CMNA matrix is 5x5 with a sparseness of 40.0%.   ◆

The compactions have one extra complication. In the above discussion, each compaction was introduced by exactly one element. However, if several of these elements are interconnected, the compactions have to be carried out subsequently and the order in which they are performed is of great importance. This will be discussed more thoroughly in the next subsection. Here, it can already be stated that a special datastructure is required to keep track of all the compactions and substitutions, both for the rows and for the columns, and that the stamps of the elements can be more complicated than for the MNA method.

4.3.1.4. Comparison of the circuit equation formulations
The three circuit equation formulations described above, the tableau approach, the MNA method and the CMNA method, are now compared with respect to matrix size, matrix sparseness and possible introduction of term cancellations for the above examples. The possible term cancellations are measured by the number of symbols appearing on more than one position in the matrix. The results for the active RC filter of Fig. 4.7 (with ideal opamps) and for the CMOS OTA of Fig. 4.8 are summarized in the following tables :

RC filter	tableau	MNA	CMNA
matrix size	54x54	15x15	5x5
sparseness	95.6%	80.9%	56.0%
cancellations	0	12	3

CMOS OTA	tableau	MNA	CMNA
matrix size	95x95	9x9	5x5
sparseness	97.3%	64.2%	40.0%
cancellations	18	18	18

Note that matching information has been exploited for the OTA, which implies that the same symbols are used for several elements. As compared to the tableau and the MNA matrices, the CMNA matrix is clearly much smaller. A compaction from 54 over 15 to 5 equations is obtained for the active RC filter. The improvement is even more significant for the CMOS OTA: from 95 over 9 to 5 equations. Moreover, for switched-capacitor circuits, the MNA method is impractical as well due to the large matrix sizes. For example, for the Fleischer-Laker biquad of Fig. 4.10 [FLE_79], the conventional MNA method [VAN_81] requires 98 equations, whereas the CMNA method suffices with 4 equations [WAL_89], as will be shown in subsection 4.3.3 below. Clearly, for the symbolic analysis of switched-capacitor circuits, the tableau and MNA methods are out of question due to their large matrix sizes. Note that for the numerical simulation of switched-capacitor circuits with SWAP [SWA_83], several algebraic matrix compactions are carried out on the conventional MNA matrix as well, in order to reduce the matrix size and the CPU time [CLA_84, RAB_83]. The minimum matrix size is then determined by the number of state variables (integration capacitors) and outputs. Topological compaction methods for switched-capacitor circuits are presented in [FAN_80, VLA_82]. They reduce the matrix size in a similar way as the CMNA method. Of course, the sparseness of the CMNA matrix decreases together with its size, but still remains large enough for the efficient application of sparse-matrix algorithms.

With respect to term cancellations, the tableau matrix has all different symbolic entries for the active RC filter, whereas the CMNA method has three elements (G3, G9 and G11) which appear on more than one position, possibly leading to term cancellations during the determinant calculation. On the other hand, the CMNA method clearly has less possible term cancellations than the MNA method for the active RC filter. In extreme cases, the number of term cancellations can even be reduced to zero, as will be illustrated in the example in the next subsection. In addition, if matching information is exploited, such as in the OTA example, the tableau matrix also shows possible term cancellations.

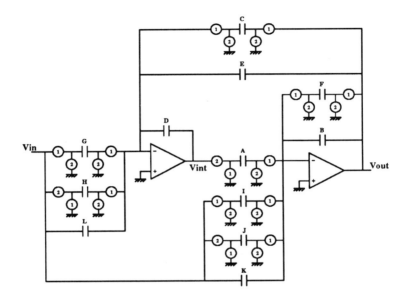

Fig. 4.10. Switched-capacitor biquad of Fleischer and Laker.

It can be concluded that the CMNA method is a most appropriate circuit equation formulation for symbolic simulation (and perhaps also for numerical linear simulation [ROH_90]). The method is generally applicable and shows an excellent compromise between matrix size and term cancellations. It clearly outperforms the MNA method and as compared to the tableau approach it has the advantage of much smaller matrix sizes. For these reasons, the CMNA method has been adopted in the ISAAC program [GIE_89c] and will be discussed more thoroughly in the next two sections for time-continuous and time-discrete circuits, respectively.

4.3.2. The set-up of s-domain CMNA equations

In the case of a time-continuous circuit (or for any particular phase of a time-discrete circuit), the s-domain CMNA equations consist of KCL equations and additional branch relations expressed in terms of node voltages and additional branch currents.

The *rows of the CMNA matrix* correspond to the applied circuit equations. According to the row compactions described in the previous subsection, the KCL equations are not applied to all nodes individually, but to the union of the terminal nodes for any dependent or independent voltage source, for any short or for any norator. This reduces the number of KCL equations and the number of rows in the CMNA matrix. At the same time, the corresponding branch current is eliminated out of the KCL equation, reducing also the number of columns. However, if some of the above elements are interconnected, the KCL equation has to applied to the union of all nodes of the interconnected elements. To keep track of this, a row-info list is set up in ISAAC, which consists of corresponding (node row-number)-pairs. Afterwards, when the element symbols are stamped into the matrix, this row-info list is inspected to retrieve the correct row number for each node. As opposed to the MNA method, several nodes can correspond to the same row number in the CMNA method. The row-info list also contains the row number which corresponds to all elements requiring an extra branch relation.

The *columns of the CMNA matrix* correspond to the variables in the circuit equations. According to the column compactions described in the previous subsection, the node voltages of any short, any current meter, any nullor and any independent voltage source, are equal or differ by a constant voltage. Hence, these node voltages can be represented by one variable only, thereby reducing the number of variables and the number of columns in the CMNA matrix. Mathematically, one node voltage is selected as variable and the other node voltage is substituted in terms of this variable, according to :

$$v_+ \Rightarrow v_- \qquad \text{or} \qquad v_- \Rightarrow v_+ \qquad\qquad (4.32)$$

or, for a voltage source, according to :

$$v_+ \Rightarrow v_- + E \qquad \text{or} \qquad v_- \Rightarrow v_+ - E \qquad\qquad (4.33)$$

For example, the first substitution $v_+ \Rightarrow v_-$ means that in all circuit equations the voltage v_+ is replaced by the variable v_-. Hence, each time v_+ is addressed, the column corresponding to v_- is accessed. In other words, the trivial branch relation of these elements which is stated explicitly in the MNA method, is used implicitly in the CMNA method and is omitted as extra equation, reducing also the number of rows.

To decide among the two possible substitutions in (4.32) or (4.33), the following rule must be followed. If n_- is grounded, v_+ must be eliminated. If n_+ is grounded, v_- must be eliminated. Otherwise, one is free to eliminate either v_+ or v_-. After the solution of the equations, the eliminated node voltage is then easily retrieved by means of the reverse (back) substitution (which are all linear).

However, a problem arises if several of the above elements are interconnected. For each of these elements, one node voltage can be substituted by the other one. But the latter node voltage can already be expressed in terms of other node voltages. Hence, in an interconnection of the above elements, the order in which the substitutions are performed is of great importance. Intuitively, a minimum set of node voltages is obtained only if the substitution process extends gradually out of one starting node.

The following algorithm finds all the necessary substitutions in a correct order. Note that there is not a unique solution. Several orders can be correct. The algorithm will find one solution only. The graph of all current meters, shorts, independent voltage sources and nullors is decomposed into maximally connected subgraphs. For each subgraph, an arbitrary starting node is chosen. If the circuit's reference node (usually AC ground) is a member of the subgraph, the reference node must be taken as the starting node. The voltage of this starting node is the only variable for the whole subgraph (except for the additional right-hand-side vectors of any independent voltage sources) with one corresponding column in the CMNA matrix. At every step in the algorithm, any branch that was not yet encountered in the subgraph and that is connected to (at least) one of the already encountered nodes with its terminal t_x is used as next branch. For this branch the voltage of the other terminal t_y is expressed in terms of v_x (which can again be expressed in terms of other, already encountered node voltages). In this way, the voltage v_y is eliminated. This process is repeated untill all branches in the subgraph have been processed. Then all node voltages in each subgraph have been expressed linearly in terms of the voltage of the starting node and possibly some independent voltage source values. In the circuit equations, each node voltage of each subgraph is substituted or replaced by this linear combination. To keep track of these substitutions, a column-info list is set up in ISAAC, which consists of corresponding (node list-of-positive-column-numbers list-of-negative-column-numbers)-triples. Afterwards, when the element symbols are stamped into the matrix, this column-info list is inspected to retrieve the correct column number(s) for each node. If an element has a terminal node, whose voltage has been substituted by a linear combination of another node voltage and some independent voltage source values, the stamp of this element is filled in each column corresponding to all variables in this linear combination. As opposed to the MNA method, the same element can be stamped in more than two columns in the CMNA method. Also, depending on the orientation of the voltage sources, this stamp can have a plus sign or a minus sign. This explains the two lists-of-column-numbers for each node in the column-info list. The column-info list also contains the column number which corresponds to all elements requiring an extra branch current and the column number which corresponds to all independent sources (for the right-hand-side vector).

The above compactions and substitutions are now illustrated in the following example.

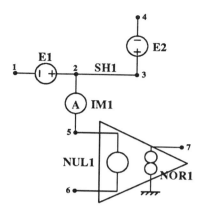

Fig. 4.11. Subcircuit to illustrate the substitutions in the CMNA method.

Example

Consider the hypothetical subcircuit of Fig. 4.11. This subcircuit is not grounded. Hence, an arbitrary starting node, for example node 1, is selected. Starting from node 1, all branches can for example be encountered in the following order :

$$E_1, \ IM_1, \ SH_1, \ E_2, \ NUL_1 \tag{4.34}$$

The resulting column substitutions are then :

$$
\begin{aligned}
v_2 &\Rightarrow v_1 + E_1 \\
v_5 &\Rightarrow v_2 \Rightarrow v_1 + E_1 \\
v_3 &\Rightarrow v_2 \Rightarrow v_1 + E_1 \\
v_6 &\Rightarrow v_5 \Rightarrow v_1 + E_1 \\
v_4 &\Rightarrow v_3 - E_2 \Rightarrow v_1 + E_1 - E_2
\end{aligned}
\tag{4.35}
$$

where v_i is the voltage of node *i*. These substitutions can be interpreted as follows. If for example an element is connected to node 4 in the above example, its symbol must be stamped into three columns instead of one, namely in the column corresponding to voltage v_1, in the column corresponding to the voltage source E_1 and with a minus sign in the column corresponding to the voltage source E_2.

With respect to the row compactions, one KCL equation is written for the union of the nodes {1,2,3,4}, one KCL equation is written for node 5, one for node 6, and no KCL equation is written for node 7 since the other side of the norator NOR_1 is grounded. Only the current through the current meter IM_1 is added as extra variable.

♦

From the above example, it is clear that rows and columns are treated differently in the CMNA method. The total number of rows and columns in the CMNA method and thus the size of the CMNA matrix is given by *(n-1-p)+q*, where *n* is the number of nodes in the circuit, *p* is the number of dependent and independent voltage sources, shorts and norators, and *q* the number of current-controlled current sources, current-controlled voltage sources and voltage-controlled voltage sources.

The compactions or substitutions described above can be carried out in two different ways. One way is to set up the noncompacted MNA matrix and carry out the appropriate symbolic add and subtract manipulations on the rows and columns to obtain the compacted CMNA matrix. However, this requires CPU time consuming symbolic manipulations and a larger computer memory area for the initial MNA matrix. In ISAAC, the other way is adopted. The circuit is preprocessed: the row-info and column-info lists are set up and then the CMNA matrix is generated directly by filling in the modified element stamps [GIE_90d]. The following example illustrates the CMNA method for a practical circuit.

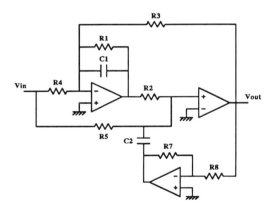

Fig. 4.12. Vogel biquad.

Example
Consider the Vogel biquad [LIB_88] of Fig. 4.12. The MNA matrix of this circuit is an *11x11*-matrix in the variables $\{v_1, v_2, v_3, v_4, v_5, v_6, v_7, i_1, i_2, i_3, i_4\}$ and each element causes term cancellations [GIE_90d]. On the other hand, the CMNA equations for the same biquad are given by :

$$
\begin{bmatrix}
-G1-sC1 & 0 & -G3 \\
-G2 & -sC2 & 0 \\
0 & -G7 & -G8
\end{bmatrix}
\begin{bmatrix}
v_3 \\
v_5 \\
v_7
\end{bmatrix}
=
\begin{bmatrix}
G4 \\
G5 \\
0
\end{bmatrix}
v_{in}
\tag{4.36}
$$

The difference in matrix size for this simple example is already remarkable, and the size of the matrix is a very important parameter for the efficiency of a symbolic equation solution algorithm. In addition, the number of term cancellations in the CMNA formulation is reduced to zero for this example. This demonstrates why the CMNA formulation is used in the ISAAC program. ◆

In this subsection, the set-up of the CMNA equations for time-continuous circuits has been discussed. The same formulation can now be applied to time-discrete circuits as well.

4.3.3. The set-up of z-domain CMNA equations

In the case of an ideal time-discrete circuit, z-domain CMNA equations can be constructed in a similar way as for time-continuous circuits. In this case, the equations are charge conservation equations and additional branch relations. If the same compactions as described before for time-continuous circuits are now applied to every phase k, then the z-domain CMNA equations can be denoted as [WAL_89] :

$$
\begin{bmatrix}
A_1 & 0 & -E_1 z^{-1} & B_1 & 0 & & 0 \\
-E_2 & A_2 & 0 & 0 & B_2 & & 0 \\
\cdots & \cdots & & & & \cdots & \\
0 & 0 & -E_N & A_N & 0 & 0 & B_N \\
C_1 & 0 & & 0 & D_1 & 0 & 0 \\
0 & C_2 & & 0 & 0 & D_2 & 0 \\
& & \cdots & & & \cdots & \\
0 & 0 & & C_N & 0 & 0 & D_N
\end{bmatrix}
\begin{bmatrix}
\underline{v}_1 \\
\underline{v}_2 \\
\cdot\cdot \\
\underline{v}_N \\
\underline{q}_1 \\
\underline{q}_2 \\
\cdot\cdot \\
\underline{q}_N
\end{bmatrix}
=
\begin{bmatrix}
\underline{Q}_1 \\
\underline{Q}_2 \\
\cdot\cdot \\
\underline{Q}_N \\
\underline{F}_1 \\
\underline{F}_2 \\
\cdot\cdot \\
\underline{F}_N
\end{bmatrix}
\tag{4.37}
$$

where $\underline{v}_k$ is the vector of node voltages and $\underline{q}_k$ the vector of additional branch charges at every phase k (k ranging from 1 to N with N being the number of phases in the circuit). $\underline{Q}_k$ is the charge source vector and $\underline{F}_k$ the voltage source vector at every phase k. The z-domain CMNA matrix is constructed as follows. First, the s-domain CMNA matrices of all ideal phases (with zero time constants) are constructed. The submatrices A_k, B_k, C_k, D_k, $\underline{Q}_k$ and $\underline{F}_k$ are then retrieved from the s-domain CMNA matrices and vectors of phase k by putting s equal to 1. The matrices E_k on the other hand are obtained from the matrices $E_k{}^*$ of the noncompacted MNA of phase k by performing the row compactions corresponding to phase k and the column compactions corresponding to the phase prior to phase k. More details can be found in [WAL_91]. The method is now illustrated in the following example.

Example
The z-domain CMNA equations for the Fleischer-Laker biquad of Fig. 4.10 are :

$$
\begin{bmatrix}
-(C+E) & -D & E\,z^{-1} & D\,z^{-1} \\
-(B+F) & 0 & B & A \\
E & D & -E & -D \\
B & 0 & -B & 0
\end{bmatrix}
\begin{bmatrix}
v_{out1} \\
v_{int1} \\
v_{out2} \\
v_{int2}
\end{bmatrix}
=
$$

$$
\begin{bmatrix}
-(G+L)\,v_{in1} + (H+L)\,z^{-1}\,v_{in2} \\
-(K+L)\,v_{in1} + (K+L)\,z^{-1}\,v_{in2} \\
L\,v_{in1} - L\,v_{in2} \\
K\,v_{in1} - K\,v_{in2}
\end{bmatrix}
\tag{4.38}
$$

These 4 equations are the charge conservation equations at the two virtual ground nodes of both opamps at phases 1 and 2. The indices 1 and 2 denote the voltage at phase 1 and phase 2, respectively. As the conventional MNA method [VAN_81] requires 98 equations for this simple example, it is clear why matrix compactions have to used both for the symbolic [WAL_89] and the numerical [CLA_84, RAB_83] simulation of switched-capacitor circuits.   ◆

4.3.4. Conclusion

In this section, the set-up of linear CMNA (Compacted Modified Nodal Analysis) equations has been described, both for time-continuous and time-discrete circuits. This formulation has been compared to other circuit equation formulations. As compared to the tableau and MNA methods, much smaller matrix sizes are obtained with the CMNA method due to the presented row and column compactions, which allow the elimination of several equations and variables. As compared to the MNA method, symbolic term cancellations are reduced or even taken away. In addition, the CMNA method is generally applicable, also for switched-capacitor circuits, and the polynomial format of all matrix entries is maintained by the compactions. All these features allow an efficient solution of the symbolic network equations afterwards and make the CMNA method the most appropriate formulation for the symbolic simulator ISAAC. The solution of these equations is now discussed in the following section.

4.4. The symbolic solution of sets of linear equations

The linear CMNA equations, which describe a linear, time-invariant circuit in the frequency domain, now have to be solved symbolically to obtain the desired network function. The equation solution algorithm is described in this section. A fast, dedicated recursive determinant expansion algorithm is presented , which exploits the typical structure of the sparse CMNA equations on one hand, and returns the symbolic output in the format required by the approximation routine on the other hand. This algorithm is compared to several others from the literature and its appropriateness and efficiency are demonstrated.

Subsection 4.4.1. first summarizes the requirements of the symbolic equation solution algorithm within the ISAAC program. According to these requirements, an optimal dedicated solution algorithm is developed in subsection 4.4.2. This algorithm is then compared to others in subsection 4.4.3.

4.4.1. Requirements for the symbolic equation solution algorithm

For the selection of an efficient symbolic equation solution algorithm, several external requirements (boundary conditions) are to be considered. Indeed, the equation solution algorithm is to be used as subroutine within the ISAAC program as shown in the flowchart in the previous chapter [GIE_89c]. This implies that requirements are imposed from the input side and from the output side of this routine.

- First of all, the algorithm operates on linear equations, which are set up in the equation set-up routine. For the equation solution algorithm, it does not matter if the equations originate from a time-continuous or from a time-discrete circuit. However, for the algorithm to be efficient, it should exploit the mathematical structure of these equations. In the case of CMNA equations, the equations are compact but still relatively sparse. This sparseness has to be exploited. Furthermore, a symbolic parameter can appear in more than one position in the matrix. This causes term cancellations, even for an algorithm which is inherently cancellation-free. Hence, these cancellations must be handled efficiently. If on the other hand a tableau formulation is used, the matrix is much larger and much sparser, but each circuit parameter appears in one position only, at least if different symbols are used for different elements and if no matching information is exploited. Under these restrictive conditions, cancellation-free results can be obtained from the larger tableau formulation in combination with a cancellation-free algorithm. This, however, does not necessarily result in smaller CPU times, due to the much larger matrix sizes. For both the CMNA and the tableau formulation, the matrix entries are in fully expanded canonic SOP form :

$$\sum_i x^i \left(\sum_j \alpha_{ij} \prod_k p_{ijk} \right) \qquad (4.39)$$

where x is the complex frequency variable (s or z), α_{ij} a numerical coefficient and p_{ijk} a symbolic circuit parameter.

• Secondly, several requirements are imposed on the equation solution algorithm from the output side. In most cases, only one or two network variables have to be solved for to obtain the desired network function. Hence, this fact has to be exploited. A full inversion of the coefficient matrix clearly is out of order. Furthermore, it must be possible to pass the calculated symbolic expression to the approximation routine, which will be described in the next section. There it is shown that no efficient and reliable approximation algorithm exists yet for nested expressions. Hence, the approximation algorithm requires the output of the calculation routine to be a ratio of two fully expanded canonic SOP forms :

$$\frac{\sum_i x^i \left(\sum_j \alpha_{ij} \prod_k p_{ijk} \right)}{\sum_i x^i \left(\sum_g \beta_{ig} \prod_h p_{igh} \right)} \qquad (4.40)$$

These considerations are now used in the next subsection to choose an appropriate and efficient linear equation solution algorithm.

4.4.2. The sparse Laplace expansion algorithm with storage of minors

The equation solution algorithm has to solve a set of symbolic linear equations :

$$A_{n \times n} \cdot \underline{x}_{n \times 1} = \underline{b}_{n \times 1} \qquad (4.41)$$

Several algorithms are now compared and an efficient algorithm will be selected at the end of this subsection.

The above set of equations can be solved by means of an **elimination algorithm** (such as Gauss or Crout [KUN_86a]), which is most efficient for equations with numerical coefficients, but not for equations with symbolic coefficients. It has been shown in section 4.2 that these algorithms are not inherently cancellation-free, suffer from large intermediate results and require a lot of intermediate symbolic processing to cancel common terms and factors. Besides, unnecessary work is done since actually all variables are solved for.

A more straightforward and frequently used method for the solution of linear equations with symbolic coefficients is the direct **application of Cramer's rule**. If one is interested in one variable x_i only, Cramer's rule is most appropriate since it presents a closed-form expression for this variable x_i :

$$x_i = \frac{|C_{nxn}|}{|A_{nxn}|} \tag{4.42}$$

where the matrix C_{nxn} is obtained from A_{nxn} by substituting column i by the right-hand-side vector b_{nx1}. Hence, two determinants have to be calculated to obtain the desired variable x_i. The result has the required format (4.40), if both determinants are expanded into a canonic SOP form. Besides, these two determinants have (n-1) columns in common and many (sub)calculations can be shared.

Note also that by adding some extra elements at the input, the circuit can be turned into a closed system [CHU_75]. The requested network function is then obtained by calculating the determinant of this closed system, and sorting the result to the extra elements. This technique has already been illustrated in the tree-enumeration example of section 4.2. In this way, the calculation of only one determinant is required to obtain both the numerator and the denominator of the requested network function.

With the application of Cramer's rule, the problem is now reduced to the calculation of symbolic determinants. In section 4.2 and [SMI_76, SMI_79], **determinant expansion methods** have been shown to be most efficient for sparse matrices. Several determinant expansion algorithms will now be presented. Starting from the determinant definition and the basic expansion algorithm, the efficiency is gradually increased, ending up with the solution algorithm used in the ISAAC program.

4.4.2.1. Application of the determinant definition

According to the definition, the determinant of an *nxn*-matrix A is given by :

$$|A_{nxn}| = \sum_{\{i_1 i_2 \ldots i_n\}} (-1)^k \, a_{1i1} \, a_{2i2} \, a_{3i3} \, \cdots \, a_{nin} \tag{4.43}$$

where k is the number of inversions in the index set $\{i_1 i_2 \ldots i_n\}$ needed to bring it into the order $\{1 \ldots n\}$, and the sum is taken over all $n!$ permutations of the set $\{1 \ldots n\}$.

As a result, in fully expanded form and with general (all different) matrix entries, the determinant contains $n!$ terms. All these terms have to be generated by the determinant calculation algorithm. This requires a CPU time of at least :

$$t = O(C) \tag{4.44}$$

where C is the complexity of the result, which is measured by the number of SOP terms for a fully expanded expression. With general matrix entries, the minimum CPU time is then :

$$t = O(n!) \tag{4.45}$$

The CPU time is in the best case linear with the complexity (or the number of terms) of the result. This number of terms, however, increases with the factorial of the matrix size. This is a well-known, but dramatic result. The value of a determinant is in the general case not bounded by a polynomial function of the matrix size, and so is not the CPU time. In this sense, the calculation of a determinant is an intractable problem of the second type as identified in [GAR_79]. The first type of intractable problems are problems that are so difficult that an exponential amount of time is needed to discover a solution. The second type are problems with a solution that is not bounded by a polynomial function of the problem size.

Fortunately, this CPU time can strongly be decreased by exploiting the sparse nature of the CMNA matrix. The summation in (4.43) can be extended over those permutations only for which the corresponding product $a_{1i1} a_{2i2} \cdots a_{nin}$ is nonzero. These nonzero permutations can be determined in advance with the following recursive procedure (an adaptation of the program PERM used in SAPEC [LIB_88]). It takes as inputs the list `colnrlist`, whose ith element initially is the list of the column indices of the nonzero entries in the ith row of the matrix, and the number `level`, which counts the recursion depth. The procedure stores all nonzero permutations in `output_list`, which is a general variable. The procedure can then be denoted in pseudo-code as follows :

```
generate_nonzero_permutations(colnrlist, level)
{
var colnr, auxlist, n;
n := length(colnrlist);
UNLESS (empty sublist in colnrlist) DO
    IF (level = n)
        THEN {push colnrlist in output_list}
        ELSE
            FOR all colnr in colnrlist[level] DO
```

```
                    {auxlist := colnrlist;
                     auxlist[level] := colnr;
                     delete colnr in auxlist[level+1]..auxlist[n];
                     generate_nonzero_permutations(auxlist, level+1)};
}
```

The determinant can then be computed by means of the calculated list of nonzero permutations `output_list` and the original matrix. A similar technique is adopted in [SAN_80]. The nonzero-permutation generation algorithm is now illustrated in the following example.

Example

Consider the determinant of the following matrix :

$$
\begin{vmatrix}
A & B & 0 & 0 \\
C & D & E & 0 \\
0 & F & G & H \\
0 & 0 & I & J
\end{vmatrix}
\qquad (4.46)
$$

The input list of the nonzero-entry column numbers for each row is then :

$$
\texttt{colnrlist} = ((1\ 2)\ (1\ 2\ 3)\ (2\ 3\ 4)\ (3\ 4)) \qquad (4.47)
$$

The algorithm then returns the following list of nonzero permutations :

$$
\texttt{output\_list} = ((1\ 2\ 3\ 4)\ (1\ 2\ 4\ 3)\ (1\ 3\ 2\ 4)\ (2\ 1\ 3\ 4)\ (2\ 1\ 4\ 3)) \qquad (4.48)
$$

The resulting determinant is then given by :

$$
A\ D\ G\ J\ -\ A\ D\ I\ H\ -\ A\ F\ E\ J\ -\ C\ B\ G\ J\ +\ C\ B\ I\ H \qquad (4.49)
$$

◆

This method exploits the sparseness of the matrix as it determines the nonzero permutations only. However, the method is still suboptimal with respect to term cancellations and hence to CPU time and memory. Indeed, first all possible nonzero terms are generated, yielding a large SOP form. Then all common terms are canceled out to obtain the much smaller, final result. This also shows that all programs which use the above technique, such as the SAPEC program [LIB_88], are suboptimal.

4.4.2.2. Sparse recursive Laplace expansion of the determinant

A better tackling of term cancellations is obtained with the **Sparse recursive Laplace Expansion method, called SLE**. In this case, the determinant $|A_{nxn}|$ is developed along an arbitrary row i as :

$$|A_{nxn}| = \sum_{\substack{j=1 \\ a_{ij} \neq 0}}^{n} (-1)^{i+j} a_{ij} |M_{ij}| \qquad (4.50)$$

or along an arbitrary column j as :

$$|A_{nxn}| = \sum_{\substack{i=1 \\ a_{ij} \neq 0}}^{n} (-1)^{i+j} a_{ij} |M_{ij}| \qquad (4.51)$$

where the summation is extended over the nonzero elements a_{ij} only, and M_{ij} is the *(n-1)x(n-1)*-matrix obtained from A by removing row i and column j. The minors $|M_{ij}|$ are recursively calculated by means of (4.50) or (4.51). As compared to the previous method, a lot of CPU time and memory is saved because the term cancellations in every minor $|M_{ij}|$ are elaborated before the minor is used on a higher recursion level. This strongly reduces the increase of the intermediate number of terms since much less terms are developed which cancel afterwards.

4.4.2.3. Sparse recursive Laplace expansion with minor storage

The previous method can still be improved further on. Much work is redone if a minor is used more than once. This same minor is then also calculated more than once, unless the minor is stored after calculation. This yields the **Sparse recursive Laplace Expansion method with Memo storage of minors, called SLEM**. Every minor at all recursion levels is calculated only once. The term cancellations are elaborated and the minor is saved in a memo table. If the same minor is needed again later on during the determinant calculation, it is retrieved from this memo table. The appropriate datastructure for such a memo table is a hash table, which allows a table access time logarithmically growing with the number of table entries. The hash keys are a function of the row and column indices of the minors. This method is now illustrated in the following example.

Example

The following determinant is calculated with the SLEM method :

$$
\begin{vmatrix} A & B & 0 & 0 \\ C & D & E & 0 \\ 0 & F & G & H \\ 0 & 0 & I & J \end{vmatrix}
$$

$$
= A \begin{vmatrix} D & E & 0 \\ F & G & H \\ 0 & I & J \end{vmatrix} - C \begin{vmatrix} B & 0 & 0 \\ F & G & H \\ 0 & I & J \end{vmatrix}
$$

$$
= A \left[D \begin{vmatrix} G & H \\ I & J \end{vmatrix} - F \begin{vmatrix} E & 0 \\ I & J \end{vmatrix} \right]
$$

$$
- C \left[B \begin{vmatrix} G & H \\ I & J \end{vmatrix} - F \begin{vmatrix} 0 & 0 \\ I & J \end{vmatrix} \right]
$$

$$
= A [D (G J - H I) - F (E J)] - C [B (G J - H I)] \tag{4.52}
$$

The minor $(G J - H I)$ is used twice but is calculated only once. This minor storage technique becomes even more important for larger examples, since already calculated lower-level minors can be used during the development of higher-level minors at all levels.

All recursive determinant expansion methods with storage of minors (as presented in this subsection and in the following subsection) can now be described in pseudo-code by the following algorithm :

```
sparse_recursive_laplace_expansion_with_memo(matrix)
{
var det, i, j, a_ij, minor;
det := 0;
select row i or column j to develop determinant along;
FOR all nonzero elements a_ij in row i or column j of the matrix DO
   IF minor(a_ij) already in memo table
      THEN det := det + (-1)^(i+j)*a_ij*minor(a_ij)
      ELSE
         {remove row i and column j from matrix;
         minor :=
            sparse_recursive_laplace_expansion_with_memo(matrix);
         det := det + (-1)^(i+j)*a_ij*minor;
         store minor in memo table;
         add row i and column j to matrix};
return det;
}
```

4.4.2.4. Sparse recursive Laplace expansion along multiple rows/columns

In the above discussion, we only considered the development of a determinant along a single row or column. This can, however, be **generalized to the Laplace expansion of a determinant along multiple rows or columns at once.**

Consider the nxn-matrix A. Select m columns (m less than n) from this matrix. Consider then all possible combinations of m rows out of the original matrix A. Each combination of m rows together with the original m columns defines two submatrices: the submatrix M of order m corresponding to the chosen rows and columns, and the submatrix CM of order $(n-m)$ corresponding to the complementary rows and columns. The determinant $|A_{nxn}|$ is then given by the sum (with appropriate signs), over all possible row combinations, of the product of the determinants of these two submatrices M and CM, according to the following formula :

$$|A_{nxn}| = \Sigma \; sign \; |M_{mxm}| \; |CM_{(n-m) x (n-m)}| \qquad (4.53)$$

where $sign$ is +1 if the sum of the selected column and row numbers is even, and -1 otherwise. The same formula holds for development along m fixed rows as well. The number of subdeterminant products in (4.53) is equal to :

$$\binom{n}{m} = \frac{n!}{m! \; (n-m)!} \qquad (4.54)$$

Formula (4.53) can be applied recursively, and in combination with the minor storage technique in order to avoid the repeated calculation of small subdeterminants.

It is now investigated which recursive Laplace expansion method, along a single row/column or along multiple rows/columns, is the most efficient one. The computational efficiency of each method is measured by the number of multiplications needed to calculate the symbolic determinant. This number of multiplications can easily be derived for the method of expansion along a single row or column in combination with the minor storage technique as follows. The number of different (ixi)-minors at the $(n-i)$th recursion level in the Laplace expansion is :

$$\binom{n}{i} = \frac{n!}{i! \; (n-i)!} \qquad (4.55)$$

Each of these minors is calculated only once due to the minor storage technique. However, according to (4.50) or (4.51), each minor can be used $(n-i)$ times by multiplying it with each of the $(n-i)$ elements, that belong to the column (row) the determinant is developed along at this recursion level and that are not part of the rows (columns) which form this minor. By summing this multiplication cost for all recursion levels, the **total computational cost** is given by :

$$\sum_{i=1}^{n-1} \binom{n}{i} (n-i) \; P_i = n \sum_{i=1}^{n-1} \binom{n-1}{i} \; P_i \tag{4.56}$$

where P_i is the number of multiplications needed to determine the product of an element with the corresponding *(ixi)*-minor. If this number would be the same for all recursion levels and equal to P, then the total computational cost of the single-row/column determinant expansion method with minor storage is :

$$n \; (2^{n-1} - 1) \; P \tag{4.57}$$

It is now proven in [GEN_74] that this cost is the smallest one (in multiplications) for all recursive minor expansion methods with memo technique. This means that no expansion along multiple rows/columns can do better than the expansion along a single row/column (although they perhaps may do as good). Therefore, in ISAAC, only expansion along a single row or column is implemented [GIE_89c], according to the SLEM algorithm presented above.

4.4.2.5. Sparse recursive Laplace expansion with minor storage and row/column ordering

There is still one degree of freedom left in this SLEM method: the selection of the row i or column j the determinant is developed along. This choice has to be taken at all recursion levels. The selection can be performed with a static decision (once for all recursion levels) or with a dynamic decision (changing during the operation of the routine). Furthermore, this choice interferes with the minor storage technique. The efficiency of the minor storage is optimized if as many lower-order minors are shared as possible. However, a dynamic row/column selection mechanism can reduce this. This will now be investigated more thoroughly.

With a *static selection mechanism*, all minors of dimension p are developed along the same column $k(p)$. This is equivalent to first ordering all columns according to a certain criterion and then developing all minors along their first column. The ordering criterion might be the number of zero entries in a column. All columns are then sorted to increasing number of nonzero entries: the most sparse columns first, the most dense ones last. In this way, it is most likely that the smallest number of minors is developed and that no minors are developed which turn out to be zero afterwards. However, this last statement is only valid at the top level. At deeper levels, some rows and columns have been deleted from the matrix and the first column of the minor at hand possibly is not the most sparse one anymore. The **Sparse recursive Laplace Expansion method with static Preordering of all columns is called SLEP.**

The simplest static selection mechanism of course is to perform no ordering of the columns at all. The order of the columns is then arbitrarily given by the matrix set-up routine, and all minors are developed along their casual first column. This is the simple Sparse recursive Laplace Expansion method, called SLE. The examples below, however, will show that this technique is least efficient.

With a *dynamic selection mechanism* on the other hand, the best column is selected at runtime for each individual minor, and the minor is developed along this best column. The selection criterion might again be the number of zero entries in a column. It is also possible to exploit the duality between columns and rows, and to extend this technique to rows as well. Each minor is then developed along its column or row with the smallest number of nonzero entries. This method is called the **Double Sparse recursive Laplace Expansion method and is denoted as DSLE.** Note that it is similar to the ordering techniques in the numerical calculation of sparse determinants.

These three methods, the SLE method, the SLEP method and the DSLE method, are now illustrated for two examples. First, a simple example is provided to give an indication of and to compare the characteristics of the different methods. The results are then confirmed for a larger example.

Example 1

The following *5x5*-determinant is calculated by sparse Laplace expansion with each of the three different row/column selection mechanisms presented above :

$$\begin{vmatrix} A & 0 & 0 & B & C \\ 0 & D & E & F & G \\ H & I & 0 & 0 & J \\ K & 0 & 0 & 0 & 0 \\ 0 & L & 0 & M & 0 \end{vmatrix} \tag{4.58}$$

1. If no ordering of the columns is performed, the determinants and all minors are developed along their first column, yielding :

```
A.{D.0 + (-I).[E.(-M).0] + (-L).E.0}

+ H.{(-D).0 + (-L).(-E).B.0}

+ (-K).{(-D).0 + I.(-E).[B.0 + (-M).C] + (-L).(-E).B.J}

= (-K).{I.E.M.C + L.E.B.J}
```
$$\tag{4.59}$$

The SLE method starts developing the minors corresponding to elements A and H, but these minors turn out to be zero. This information, however, can be retrieved directly from the matrix, since these minors contain a row of zeros. Clearly, if no ordering of the rows and columns is performed, the sparseness information of the matrix is not exploited very well.

2. In the second method, the SLEP method, all columns are first ordered to increasing number of nonzero entries. This results in the following order {3,1,2,4,5}. The determinant and all minors are then developed along their first column in this list, yielding :

```
(-E).{A.[I.(-M).O + L.O]

    + (-H).L.B.O

    + K.[(-I).(B.O + (-M).C) + L.B.J]}

= (-E).[K.I.M.C + K.L.B.J]
```
$$\tag{4.60}$$

The sparseness information is now exploited at the top level: the element E, which is a single entry in a column, is factorized first. However, the problem encountered with the first method (when no column ordering is performed at all) now arises one level deeper: during the calculation of the minor of the element E, several subminors which have a row of zeros are still being developed, only returning zero at the end.

3. The sparseness information is only fully exploited if the determinant is developed both along rows and columns. In the third method, the DSLE method, the determinant and all minors are developed along the column or row with the largest number of zero entries. In the above example, the determinant is first expanded along the third column (which contains one nonzero entry), the corresponding minor is then developed along the fourth row (which again contains one nonzero entry), and so forth. This yields :

```
(-E).K.[(-I).(-C).M + L.B.J]

= (-E).K.[I.C.M + L.B.J]
```
$$\tag{4.61}$$

The same result is now obtained immediately, without developing any minors which turn out to be zero afterwards. The DSLE method clearly exploits the sparseness information in the most efficient way.

This example also shows **two important properties of the DSLE method**. First, single nonzero entries in a row or column are developed first. This yields exactly factorized results at the top level. Secondly, the DSLE method does not develop any minors which obviously are zero because of a row or column with all zero entries. The number of terms being developed grows in almost all cases monotonically to the final number of terms [SMI_79]. No intermediate expression growth is observed. The DSLE method does not develop many more terms than there are nonzero terms in the final result. Only if a minor, which is not obviously zero, turns out to be zero because of subminors with all-zero rows or columns (which occurs only occasionally when using the DSLE method) or because of term cancellations, terms are being developed which are not present in the final result. Therefore, **the DSLE method (eventually in combination with the minor storage technique) is mathematically the most attractive method.** Note that the final number of terms for the determinant of a matrix with general (all different) entries is given by the permanent of the matrix formed from the original matrix by replacing each nonzero entry by +1 (thus ignoring the actual values, only exploiting the zero/nonzero information). This permanent is 2 in the previous example.

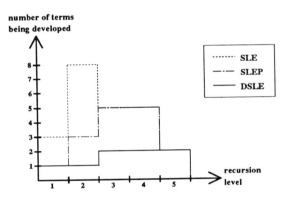

Fig. 4.13. Number of terms being developed as a function of the recursion depth by several recursive determinant expansion algorithms for a simple example.

The number of terms being developed at recursion level q can be measured by the number of times the determinant expansion routine is called for a minor with dimension $(n\text{-}q)$, where n is the size of the original matrix. For the above example, the number of terms being developed as a function of the recursion depth are then {3,8,5,5,2}, {1,3,5,5,2} and {1,1,2,2,2} for the SLE method, the SLEP method and the DSLE method, respectively. This is schematically shown in Fig. 4.13. The SLE method has a large initial growth of the number of terms. The SLEP method has a similar increase at lower recursion levels, whereas the DSLE method shows a monotonic growth of the number of terms.   ♦

Example 2

This behavior of the different algorithms is now illustrated for the calculation of the system determinant of the active RC bandpass filter taken from [STA_86] and shown in Fig. 4.14. The corresponding CMNA matrix sizes 18 by 18. The number of terms being developed as a function of the recursion depth for the SLE, the SLEP and the DSLE methods are shown in Fig. 4.15. Note the logarithmic y-axis. Only the DSLE method shows a monotonic growth of the number of terms.   ♦

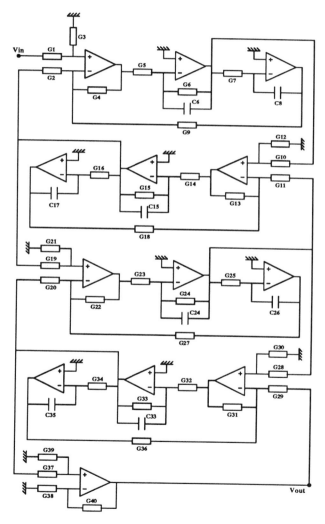

Fig. 4.14. Active RC bandpass filter.

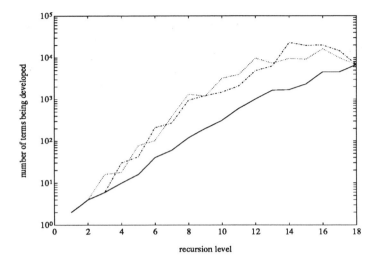

Fig. 4.15. Number of terms being developed as a function of the recursion depth by the SLE method (dotted), the SLEP method (dashed) and the DSLE method (straight) for the active RC bandpass filter of Fig. 4.14.

Since the SLE and SLEP methods initially try to develop more terms, they require a larger CPU time than the DSLE method. However, if a minor turns out to be zero, the determinant expansion routine just tracks back and no time-consuming polynomial multiplications and additions have been performed yet. Moreover, the gain in CPU time with the DSLE method is partially undone by the time required to select the best column or row for all minors at all recursion levels. Obviously, much time is lost if for each minor the number of nonzero elements in each row and column is counted again and again. To perform the selection efficiently, a special datastructure is required to keep track of the number of nonzero elements in each row and column of the (always changing) minor. Note that such an additional counting mechanism is also required if the matrix itself is stored with a sparse-matrix pointer-based datastructure, as the minor is always changing. One possibility for such a mechanism is then to store the number of nonzero entries in each row and column in a special counting list. Each time a subminor is developed, one row and column are deleted from the current minor and the counting list is updated. The best column or row is then selected by tracing this counting list. In [SMI_79], a histogram is used for the selection. Rows and columns with the same number of nonzero entries are grouped in equivalence classes, which are stored in the histogram. This histogram is dynamically updated as rows and columns are removed from or inserted into the matrix. The selection of a best row or column is then performed by choosing one member of the nonempty equivalence class with the lowest number of nonzero entries.

Also, to avoid multiple calculations of the same subminor in the sparse recursive Laplace expansion method with one of the above ordering mechanisms, the minor storage technique (memo table) can be used. However, it is important to notice that the dynamic selection of a best row or column conflicts with the minor storage technique. If each time another row or column is selected to develop a minor of dimension p along, then most propably less minors will be reused (with the corresponding increase in CPU time) than if all minors of dimension p are developed along the same row or column. On the other hand, minors with none or one nonzero element in a row or column are recognized faster and handled more efficiently, with the observed monotonic behavior as a result. These factors influence the CPU time in opposite directions. The overall efficiency of both methods, the sparse recursive Laplace expansion method with minor storage and with static or dynamic row/column selection, will be compared more thoroughly in the next subsection.

It is also important to notice that the application of Cramer's rule implies the calculation of two determinants which have $(n-1)$ columns in common. This allows to share subresults among both calculations if the column corresponding to the variable of interest x_i is excluded from the minor storage mechanism. A possible method is then to develop both the numerator and denominator determinant at top level along the column corresponding to x_i, and to apply the above techniques (minor storage with static or dynamic row/column selection) on all minors (top and lower levels), which are then common to the numerator and the denominator determinant. This is equivalent to changing the ordering criterion slightly, such that the column corresponding to x_i has the highest priority.

Remark

For the efficiency of symbolic analysis, it is important that each circuit is reduced to its minimal form. If there are nonobservable or noncontrollable nodes in the circuit, then there can be coinciding zeros and poles in the network function. This can for instance be the case if the circuit contains a subcircuit which is not affected by the input excitation. This subcircuit then strongly increases the CPU time and the complexity of the resulting expressions, as it results in exactly cancelling zeros and poles between the numerator and the denominator of the network function. Another example is a symmetrical circuit, which can be analyzed by analyzing only one of its two symmetrical subcircuits. So, if the determinant routine expands the expressions, the expressions first have to be factorized again before these pole-zero cancellations can be carried out. In addition, these pole-zero cancellations have to be carried out before the expressions are approximated, since the approximation may discard small terms which are needed for the polynomial factorization, resulting in unfactorizable expressions. Clearly, the cancellation of coinciding zeros and poles and symbolic approximation do not commute. Therefore, it is important to detect such situations in advance before the network function is calculated. This can be accomplished by inspecting the matrix of the circuit equations, especially if for instance a state equation formulation is used, or by topological checking of the circuit schematic.

Conclusion

In this subsection, several algorithms for the solution of sets of symbolic linear equations have been presented. The efficiency of these algorithms will be discussed more thoroughly in the next subsection. The sparse recursive Laplace expansion method with minor storage and dynamic row/column selection combines all efficient mechanisms to exploit the sparseness of the equations and to reduce the CPU time for the calculation of expanded results. In the next subsection, its performance will also be compared to other solution algorithms presented in the literature and it will be shown that up till now this method is the best one for the problem at hand, the calculation of determinants within the ISAAC program.

4.4.3. Comparison of the efficiency of symbolic solution algorithms

In this subsection, first the efficiency of the recursive determinant expansion algorithms presented in the previous subsection is discussed. The most efficient algorithm, which is adopted in the ISAAC program [GIE_89c], is then compared to other symbolic solution algorithms presented in the literature.

4.4.3.1. Comparison of recursive determinant expansion algorithms

A lower bound and thus the best-case result for the required CPU time to calculate a determinant in symbolic form is given by :

$$t = O(C) \tag{4.62}$$

where C is the complexity of the result. For an expanded canonic SOP form, this complexity is given by the number of terms. For nested expressions, other complexity measures are needed. An expression for the number of terms as a function of the network size is thus an indication of the minimum CPU time required to generate the symbolic result in expanded form since all these terms simply have to be generated. An efficient solution algorithm then consumes a CPU time, which grows linearly with the number of terms. **This behavior as a function of the number of terms is one efficiency measure for a symbolic equation solution algorithm.** This is now investigated for the SLEM and the DSLEM methods for the following two extreme examples.

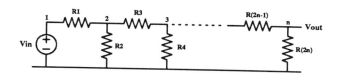

Fig. 4.16. Resistive ladder network as example of a sparse network.

Example 1

The first example is a resistive ladder network, shown in Fig. 4.16. This is a very sparse example since each node is only resistively connected with its predecessor and its successor node and the ground. If we denote the number of ladder sections with n, then the number of resistors is $2n$. The number of canonic SOP terms in the transfer function of this ladder filter is 1 for the numerator and F(2n) for the denominator, where F(i) is the ith Fibonacci number. Hence, the number of terms and the minimum CPU time increase with the size of the network as :

$$t = O\left(\left(\frac{1+\sqrt{5}}{2}\right)^{2n}\right) \tag{4.63}$$

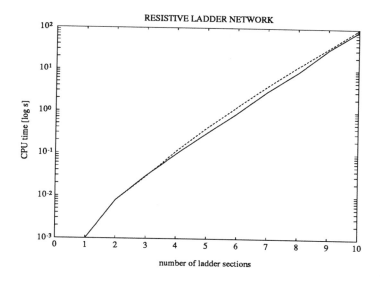

Fig. 4.17. CPU time as a function of the size of the resistive ladder network for the DSLEM algorithm (straight line) and for the SLEM algorithm (dashed line).

Fig. 4.17 illustrates this for the SLEM and the DSLEM algorithms. The CPU time indeed increases linearly with the number of terms, which increases exponentially with the size of the network according to (4.63).   ♦

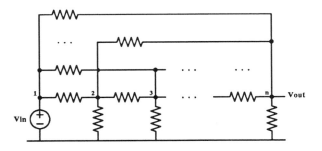

Fig. 4.18. Completely connected resistor network as example of a dense network.

Example 2

The second example is a completely connected resistor network, shown in Fig. 4.18. This is an absolutely dense example since each node is resistively connected with all other nodes. If we denote the number of nodes (without the ground node) by n, then the number of resistors is given by :

$$(n-1) \; + \; \frac{n.(n-1)}{2} \tag{4.64}$$

The number of canonic SOP terms in the transfer function of this ladder filter is given by $(n+1)^{(n-2)}$ for the numerator and $2(n+1)^{(n-2)}$ for the denominator. Hence, the number of terms and the minimum CPU time increase with the size of the network as :

$$t \; = \; O(n^n) \tag{4.65}$$

Fig. 4.19 illustrates this for the SLEM and the DSLEM algorithms. The CPU time indeed increases linearly with the number of terms, which increases more than exponentially with the size of the network according to (4.65).   ♦

A second efficiency measure for a symbolic solution algorithm is given by the number of term cancellations. The CMNA formulation possibly leads to term cancellations, whereas the approximation algorithm requires a canonic (cancellation-free) SOP form. Hence, an important measure is how efficient the algorithm treats these cancellations, how many terms are generated which cancel afterwards, or how many term cancellations are carried out by the algorithm as a function of the circuit size since these cancellations increase the total CPU time.

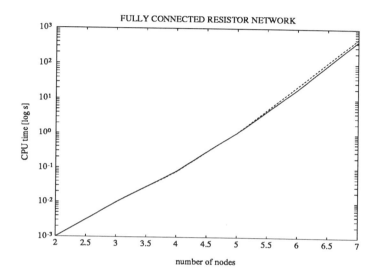

Fig. 4.19. CPU time as a function of the size of the completely connected resistor network for the DSLEM algorithm (straight line) and for the SLEM algorithm (dashed line).

The number of term cancellations of course depends on the solution algorithm. The ratio of the number of term cancellations to the final number of terms as a function of the network size when applying the nonzero-permutation generation method, the SLE method and the SLEM method are shown in Fig. 4.20 for the resistive ladder network of Fig. 4.16.

Clearly, the SLEM method is the most efficient one with respect to term cancellations for this sparse example. The number of cancellations for the nonzero-permutation generation method keeps on increasing with the circuit size, whereas it reaches a constant level (related to the final number of terms) for the SLE and the SLEM methods. This level, however, is smaller for the SLEM method than for the SLE method because of the minor storage technique. Similar results can be derived for more dense examples as well. As a result, it can be concluded that the SLEM method is the best of these methods for the generation of expanded canonic SOP forms.

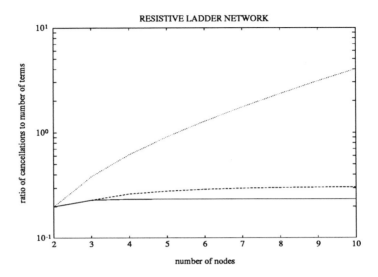

Fig. 4.20. Ratio of the number of term cancellations to the final number of terms as a function of the circuit size for the nonzero-permutation generation method(dotted), the SLE method (dashed) and the SLEM method (straight).

The **influence of the row/column ordering mechanism on these algorithms** is now investigated. To this end, we empirically compare the efficiency of the following algorithms for some practical examples: sparse Laplace expansion without memo table (SLE), sparse Laplace expansion with memo table (SLEM), sparse Laplace expansion with memo table and static preordering (SLEMP) and sparse Laplace expansion with memo table and dynamic row/column selection (DSLEM). The results are summarized in the following invocation count tables. The first column gives the *recursion depth*. The (sub)matrices at recursion level p then have the dimensions $(n+1-p) \times (n+1-p)$. The columns labeled *counts* give the invocation count results of the determinant expansion routine. This count is increased at recursion level p each time the determinant routine is entered with a (sub)matrix of dimension $(n+1-p) \times (n+1-p)$, i.e. each time a nonzero entry is found at the next higher level and hence the corresponding minor has to be calculated. The columns labeled *memo* give the number of minors retrieved from the memo table as a function of the recursion level. Finally, the total number of invocation counts, the total number of minors retrieved from the memo table and between brackets the total number of entries in the memo table, the total number of times that a minor turns out to be zero (*zero count*), the total number of terms in the final result and the CPU time (in seconds) on a Texas Instruments EXPLORER II Plus are given. The total number of invocation counts is a measure for the number of multiplications needed to calculate the result. A high zero count indicates that the sparseness is exploited badly. The CPU time is given to provide a relative indication of the performance of the different methods.

Examples

The first example is a resistive ladder network, as shown in Fig. 4.18, with 10 sections. The second example is the bandpass filter shown in Fig. 4.14 [STA_86]. The invocation count results for these two examples for the above four algorithms are summarized in the two tables below.

ladder	SLE		SLEM		SLEMP		DSLEM	
recursion depth	counts	memo	counts	memo	counts	memo	counts	memo
1	1	0	1	0	1	0	1	0
2	2	0	2	0	2	0	2	0
3	4	0	4	1	4	0	3	1
4	8	0	6	0	8	2	3	0
5	16	0	12	3	12	3	5	1
6	28	0	15	3	15	3	6	2
7	48	0	18	3	18	3	6	2
8	80	0	21	3	21	3	6	2
9	132	0	24	3	24	3	6	2
10	150	0	13	0	13	0	6	0
total count	469		116		118		44	
memo count	0		16		17		10	
[memo size]	[0]		[86]		[87]		[27]	
zero count	218		71		70		0	
nr of terms	10946		10946		10946		10946	
CPU time [s]	111		89		137		86	

Note that the results in these tables are for the calculation of the denominators of the signal transfer function only. The calculation of the numerator takes advantage of the minors already calculated and stored during the denominator calculation and requires less than 1 s for both examples.

The following conclusions can be drawn from the above examples. The SLE method has the largest invocation counts and most of the minors turn out to be zero afterwards. As a result, this method requires large CPU times. The introduction of the memo table in the other three methods clearly reduces the invocation counts. The SLEM and the SLEMP methods show the predicted growth of the intermediate number of terms, whereas the DSLEM method shows a more monotonic behavior. The SLEM and the SLEMP methods also exploit the sparseness much less efficient than the DSLEM method, since they develop many minors which turn out to be zero resulting in a high zero count. The static preordering clearly shows no advantage over the simple SLEM method for the above examples and even results in larger CPU times than the simple SLE method. In addition, if the memo counts are compared to the total invocation counts and if the total size of the memo table is considered, then the DSLEM method apparently uses the memo table as efficient as the other methods and requires much less memo storage. On the other hand, the DSLEM method consumes extra CPU time to detect the best row or column to develop each minor along. This drawback to some extent compensates the advantages of the DSLEM method. The net result is that the DSLEM method in its current implementation requires a CPU time slightly smaller than the SLEM method. In addition, the DSLEM method is mathematically also the most powerful and most attractive algorithm and its efficiency can still be increased by improving the implementation of the dynamic row/column selection (which is a critical part of the algorithm).   ♦

filter	SLE		SLEM		SLEMP		DSLEM	
recursion depth	counts	memo	counts	memo	counts	memo	counts	memo
1	1	0	1	0	1	0	1	0
2	2	0	2	0	2	0	2	0
3	4	0	4	0	4	0	2	0
4	14	0	14	0	6	0	2	0
5	16	0	16	2	12	0	3	0
6	26	0	22	0	18	0	4	0
7	36	0	30	0	36	0	4	0
8	54	0	42	0	48	0	5	1
9	176	0	136	0	72	0	5	0
10	164	0	124	10	96	0	7	0
11	178	0	119	0	120	0	8	0
12	226	0	150	0	180	0	10	3
13	535	0	356	5	216	0	8	0
14	412	0	268	31	288	0	11	0
15	540	0	297	0	384	0	11	1
16	252	0	114	39	288	96	9	5
17	374	0	95	3	120	48	7	0
18	195	0	32	0	30	0	7	0
total count	3205		1822		1921		106	
memo count	0		90		144		10	
[memo size]	[0]		[1699]		[1746]		[88]	
zero count	2859		1681		1551		4	
nr of terms	6840		6840		6840		6840	
CPU time [s]	167		158		184		133	

Therefore, it can be concluded that **the DSLEM algorithm is the best recursive determinant expansion algorithm if an expanded expression is required**. For this, the DSLEM algorithm is used in the present version of ISAAC [GIE_89c]. In the next subsection, the DSLEM algorithm is now compared to other symbolic solution algorithms presented in the literature.

4.4.3.2. Comparison to other symbolic solution algorithms

The DSLEM method is compared here to the symbolic solution algorithms used in other determinant-based symbolic analysis programs.

In SAPEC [LIB_88], the determinant of the MNA matrix of the circuit is calculated according to the determinant definition by means of a nonzero-permutation generation method, similar to the one presented in subsection 4.4.2.1. There, it has also been shown that this method is less efficient than the recursive Laplace expansion methods.

A similar approach is used in CASNA [DEL_89]. In this program, the circuit equation formulation of Sannuti and Puri [SAN_80] is used which results in a matrix with symbolic elements on the diagonal only. The determinant of this Sannuti-Puri matrix is then calculated according to the determinant definition by means of an efficient nonzero-permutation generation method, which is the algorithm presented in 4.4.2.1 extended with a technique which is similar to the memo storage concept. In this way, this method becomes comparable to the SLEM algorithm, but it still exploits the sparseness not as good as the DSLEM algorithm. In [DEL_89] it is even shown that the straightforward application of this method to the Sannuti-Puri matrix is more efficient than the original determinant calculation formula presented by Sannuti and Puri themselves [SAN_80].

Clearly, as shown above and as recognized in [SMI_79] and [SAS_82], **recursive determinant expansion methods are the most efficient algorithms for the calculation of sparse symbolic determinants in expanded form**. In [SMI_81b], however, another determinant expansion algorithm, the FDSLEM algorithm, is presented. It is an extension of the DSLEM algorithm, which returns compact symbolic expressions in nested and possibly factorized format. It is used in the NETFORM program which applies a modified tableau matrix in which the KCL and KVL equations are made as sparse as possible by introducing extra variables [SMI_81a]. In the FDSLEM algorithm, the result is expressed in nested form and common subexpressions are referenced by a proper symbol, which represents the expression further on. The value of this symbol is calculated only once. Furthermore, a factorization is performed at each level if possible. The facilities of this algorithm can be extended with a power series truncation facility [SMI_78b]. This is, however, only a primitive form of approximation, as will be discussed in the next section 4.5.

Example
The following example illustrates the FDSLEM algorithm :

$$
\begin{vmatrix}
A1 & A2 & 0 & 0 \\
B1 & B2 & 0 & 0 \\
0 & C2 & C3 & C4 \\
0 & 0 & D3 & D4
\end{vmatrix}
$$

$$
= A1 \begin{vmatrix}
B2 & 0 & 0 \\
C2 & C3 & C4 \\
0 & D3 & D4
\end{vmatrix} - A2 \begin{vmatrix}
B1 & 0 & 0 \\
0 & C3 & C4 \\
0 & D3 & D4
\end{vmatrix}
$$

$$
= A1\ B2 \begin{vmatrix}
C3 & C4 \\
D3 & D4
\end{vmatrix} - A2\ B1 \begin{vmatrix}
C3 & C4 \\
D3 & D4
\end{vmatrix}
$$

$$
= A1\ B2\ CD34 - A2\ B1\ CD34
$$

$$
= AB12 * CD34
$$

$$
= ABCD1234 \tag{4.66}
$$

Note the nested representation which allows for a complete factorization in the above example. ♦

The factored/nested representation of the result has two main advantages: the symbolic result is more compact and the CPU time needed to obtain the result is smaller. However, the nested expression is only useful if no cancellations occur in the final nested result anymore. The value of the determinant calculated with the FDSLEM algorithm is given as a nested expression with operators {/,*,-,+} in which all subexpressions are irreducible if all nonzero matrix entries are replaced by general values. This reveals the importance of the NETFORM requirements that each circuit element is represented by a different symbol and that each symbol appears only in one position in the matrix. As a result, this method cannot exploit matching information between elements. If the matching would be introduced by a postprocessor, this would require (at least a partial) expansion of the expressions. Furthermore, as will be shown in the next section 4.5, no practical approximation algorithm exists yet for nested expressions, only for expanded expressions. Finally, the modified tableau equations used in the NETFORM program are much larger and sparser than the CMNA equations. Since the expressions have to be expanded for the approximation anyway, it can be concluded that the FDSLEM technique, which can be optimal on its own [SMI_81b], is slower than the application of the DSLEM algorithm on the CMNA equations. The selection of the most optimal algorithm clearly is related to the formulation of the circuit equations and the requested format of the output expressions. **For the requirements imposed on ISAAC, the DSLEM algorithm in combination with the CMNA equations is the best solution** and hence is adopted in the program. This combination outperforms all other symbolic determinant algorithms published up till now for the given requirements. In section 4.6 it is also shown experimentally that this method is at least as good as topological methods as well.

4.4.4. Conclusion

In this section, the solution of sets of symbolic linear equations has been discussed. A dedicated and fast solution algorithm has been developed, which is adapted to the mathematical structure of the sparse CMNA equations and to the required canonic SOP format of the output expressions. This format is imposed by the expression approximation method, which will be described in the next section. The equation solution algorithm is based on a double sparse recursive Laplace expansion of the determinant with storage of minors. This algorithm has been shown to outperform all other determinant algorithms for the given requirements.

4.5. Symbolic expression approximation

The network functions, generated with the symbolic determinant expansion algorithm of the previous section, are exact, fully expanded expressions. For practical analog integrated circuits, these expressions are usually lengthy and difficult to interpret. Therefore, in this section, a heuristic approximation algorithm is presented, which approximates (or prunes) the calculated network function up to a user-defined error, based on the (order of) magnitude of the elements.

In the first subsection 4.5.1, a general introduction to symbolic expression approximation is given. The requirements imposed on the approximation routine in the ISAAC program are described and the need for an expanded SOP form is shown. An appropriate error definition, which adequately handles large terms with opposite sign, is then presented in subsection 4.5.2. This leads to the formulation of an efficient heuristic symbolic approximation algorithm in subsection 4.5.3. This algorithm combines a local and a global strategy to increase the efficiency. The algorithm is then compared to the literature in subsection 4.5.4.

4.5.1. Introduction to symbolic expression approximation

In this subsection, first symbolic expression approximation is defined and the requirements for a symbolic approximation routine within the ISAAC program [GIE_89c] are indicated. Then, it is shown that to satisfy these requirements the expressions have to be in expanded format.

4.5.1.1. Requirements for the symbolic approximation routine

Approximation theory is an important research domain in mathematics and computer sciences and especially in numerical analysis [CHE_66, POW_81]. In general, the problem can be defined as follows. Given a multivariate function $g(\underline{x})$, find an approximation $h(\underline{x})$ from a given set Υ (e.g. the polynomials of degree at most n) to the function $g(\underline{x})$ over the interval $[\underline{a}, \underline{b}]$ such that the error measured according to some norm (e.g. the least-squares norm) is minimal or within a given bound over this interval. This is schematically shown in Fig. 4.21 for the approximation of the univariate function $g(x)$ by the function $h(x)$ over the interval $[a, b]$. Many theoretical results exist with respect to the existence and the uniqueness of best approximations, and many methods have been published to determine or calculate approximations (among which spline functions are very powerful for computer calculations nowadays) [CHE_66, POW_81].

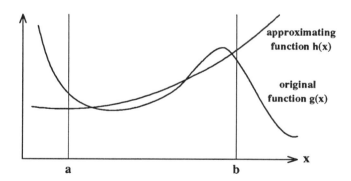

Fig. 4.21. Approximation of the univariate function $g(x)$ by the function $h(x)$ over the interval $[a, b]$.

All of the methods in [CHE_66, POW_81], however, deal with the approximation of (mostly univariate) functions with numerical coefficients. No publications have been found which consider the approximation of rational or polynomial functions with symbolic coefficients. This problem is tackled in this section. **The aim of approximation in the case of symbolic expressions is to reduce the complexity and to improve the interpretability of the expressions,** whereas the objective in the case of numerical functions is to replace a complicated function (or some data) by one that can be handled more efficiently by computers. For the symbolic case, the approximation problem can then be redefined as follows. Given a symbolic polynomial function $g(\underline{x})$ of degree n, find an approximating polynomial $h(\underline{x})$ of degree at most n and whose symbolic coefficients are composed of terms drawn from the terms of the original polynomial $g(\underline{x})$, over the interval $[\underline{a}, \underline{b}]$ such that the complexity of the polynomial is reduced as much as possible while the approximation error measured according to some suitable norm is below a maximum value ε_{max} over this interval. It is thus trading off expression accuracy (error) against expression simplicity (complexity).

In this text, **a heuristic approach towards the approximation of symbolic expressions, resulting from network functions, is followed**, which has resulted in good results for many practical examples and which is based on the practice of designers when they manually derive expressions. Heuristically, approximating a symbolic expression can be viewed as pruning insignificant terms based on the magnitude of these terms. The general idea is to discard smaller terms against larger terms in order to obtain a much shorter expression, which still describes the circuit behavior accurately enough. It is thus a trade-off between accuracy (error) and simplicity (complexity). This trade-off is visualized in an ε-C-plot, such as Fig. 4.22, *where the formula complexity C is plotted versus the error ε.* Heuristic measures for the complexity and the error of canonic SOP forms will be described in the next subsection 4.5.2.

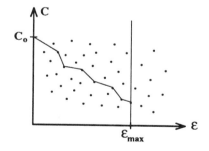

Fig. 4.22. Complexity C of an expression as a function of the approximation error ε, and trace of an approximation algorithm.

Note also that any noncyclic expression can be represented as a tree. The nodes then correspond to operators (/, *, +) and the leaves to numbers and symbols. This tree representation is particularly useful for nested formulas. With this tree idea in mind, approximating an expression can now be viewed as pruning leaf cells or whole subtrees from the original tree and/or rearranging the original tree according to some strategy. In the case of expressions in expanded canonic SOP format, however, the only possibility is to remove leaf cells or whole subtrees from the original tree. The need for such an expanded format will be shown in the next subsection.

If now an arbitrary number of randomly chosen terms are removed from the original expression, a new expression is obtained with a lower complexity and with another error. This new expression is an approximation of the original expression and can be represented as a point in the ε-C-plot. As a result, many approximations exist for a given formula, but not all approximations are useful. A reasonable approximation yields the formula with the smallest complexity for a given error or the formula with the smallest error for a given complexity.

Any approximation algorithm then proceeds by repeatedly removing (one or more) terms from the expanded expression. At each step a new approximating formula is obtained, which corresponds to a point in the ε-C-plot. If the points of the subsequent approximations are interconnected by lines, the whole approximation process is represented by a path in the ε-C-plot, also called the *trace of the approximation method*. An example of such a trace is also shown in Fig. 4.22. Different approximation strategies correspond to different ways to select how many and which terms to remove in each step, and hence have different traces in the ε-C-plot. Note that the trace of a particular approximation strategy also depends on the definitions of complexity and error used.

The **development of an expression approximation routine within the ISAAC program is now subjected to the following boundary conditions** :
- first, the user must be able to control the error (or complexity) of the approximation. Analog designers prefer to supply a maximal error ε_{max} not to be exceeded during the approximation. Moreover, they want to decrease or increase this error gradually to obtain more accurate or more simple formulas. Hence, the error percentage can continuously range from 0% to 100%. Indeed, first-order formulas (with a large error) are appropriate to gain insight into a circuit's behavior and at the initial design of a circuit. More accurate formulas are needed for the fine-tuning and verification of the circuit.
- secondly, the error supplied by the user is not to be interpreted as a purely mathematical parameter. *The final goal is to provide the user with the expression with the largest information content for the given error*. This has several implications. Consider for instance an expression which contains a subset of terms which are all of the same magnitude. If all terms would be dropped, the error would be exceeded. By neglecting some of the terms only, the error can be reached exactly. However, this provides the user with confusing results since it is rather arbitrarily decided then which terms are present and which are removed, though all terms are of the same magnitude. Hence, to provide the user with the expression with the largest information content within the given error, all terms have to be included for this example. In general, if a term of a given magnitude is present, then all terms of this magnitude and certainly all terms of a higher magnitude have

to be present (even if this leads to an error value below the user-supplied maximum value). Also, if a term is present, it has to be present with the right numerical coefficient. This condition may not be fulfilled while approximating nested expressions.

- thirdly, the magnitude of an element is determined by its value in the input file or in the magnitude file, is taken from a built-in default value, or is supplied by the user, as already explained in the previous chapter. These values can be exact values (if the design has already been completed), or can be estimated/default values (if the design still has to be performed). In the latter case, the final values (after the design) can substantially differ from the estimated/default values. But even in the first case, the values can differ from the nominal values due to processing tolerances. As a result, the approximation algorithm may not be too sensitive to the actual values of the circuit parameters and should effectively cope with parameter variations and tolerances.

- fourthly, the approximation also depends on the frequency, since for example parasitic capacitances are important at higher frequencies only. Hence, it should be possible to carry out the approximation over the whole frequency range (to include both DC and high-frequency effects in one and the same formula), over a limited frequency range or for a specific value of the frequency. This center frequency is then provided by the user and can be selected on a plot (obtained from numerically evaluating the exact symbolic expression of type 3).

- a final requirement is that the approximation result for a given maximum error and a given set of magnitude values has to be unique. It may not depend on, say, the ordering of the terms in the original expression or the numbering of the circuit nodes.

With these requirements in mind, an appropriate error definition and approximation algorithm will be derived for the approximation of symbolic network functions. But before doing so, the required format of the network functions is discussed in light of the above conditions.

4.5.1.2. Format of the expressions needed for the approximation algorithm

Obviously, the most compact form for a network function is the nested format. However, the nesting poses serious problems **if the expression has to be approximated over the whole frequency range**. In this case, the approximation is carried out coefficient per coefficient of the frequency variable ξ (σ or ζ). Consider for example the following rational function in σ :

$$H(s) = \frac{N(s)}{D(s)} = \frac{A_0 + A_1 s + A_2 s^2 + A_3 s^3 + \ldots}{B_0 + B_1 s + B_2 s^2 + B_3 s^3 + \ldots} \tag{4.67}$$

The numerator and the denominator of this rational function are approximated separately. The resulting relative error on the magnitude of the transfer function is then limited by the sum of the relative errors on the numerator and the denominator separately. Consider now one polynomial, denoting either the numerator or the denominator :

$$P(s) = A_0 + A_1 s + A_2 s^2 + A_3 s^3 + \ldots \qquad (4.68)$$

All coefficients A_i of this polynomial are then approximated individually up till the maximum user-supplied error ε_{max}. The resulting relative error on the magnitude of the polynomial $P(s)$ at the angular frequency ω is then given by :

$$\frac{\Delta |P|}{|P|} = \sum_i (-1)^{[i/2]} \frac{f(P) \, A_i \, \omega^i}{|P|^2} \, \delta A_i \qquad (4.69)$$

where $f(P)$ is equal to the real part of P for i even and to the imaginary part of P for i odd, δA_i is the effective relative error on the coefficient A_i (between $-\varepsilon_{max}$ and ε_{max}), and $[a]$ is the integer part of number a. Formula (4.69) clearly shows that if all coefficients A_i are uniformly approximated with the same relative error $\delta A_i = \delta A$, the relative error on the magnitude of the polynomial P is also given by $\delta |P| = \delta A$ over the whole frequency range. In that case, the relative error on the phase of the polynomial as well as on the roots is zero, as these are all determined by a ratio of polynomial coefficients.

In general, however, the effective errors obtained after the approximation can differ for the individual coefficients A_i. This will then effect the relative error on the magnitude of the polynomial according to (4.69). In addition, it will introduce an error on the roots of the polynomial as well. Such a situation can occur in practice, for example because of the second requirement given in the previous subsection. In ISAAC, priority is given to providing the user with the expression with the largest information content over providing him an expression with exactly the supplied error value. As a result, the individual coefficients can have different error values. Experiences with many practical circuits have shown that this has usually only a small influence on the magnitude response. It can, however, sometimes result in rather large errors on the roots of the polynomial, even with a small error tolerance on the individual coefficients. Consider for example a single root given by $(A_0 + A_1 s)$. The relative root variation δz is then to first order given by the difference between the two relative errors on the individual coefficients A_0 and A_1 :

$$\frac{\Delta z}{z} = \frac{\Delta A_0}{A_0} - \frac{\Delta A_1}{A_1} \qquad (4.70)$$

Both effective relative errors can be of the same sign or of opposite sign, but they are both smaller than ε_{max} in absolute value. Hence, the worst-case relative deviation of the root is bounded by :

$$\left| \frac{\Delta z}{z} \right| \leq 2\ \varepsilon_{max} \tag{4.71}$$

An extreme example can be found for two real poles which are close to each other in the exact expression and which may become complex in the approximated expression. However, **in most practical cases and especially for the magnitude response, the method of individually approximating all coefficients has proved to work very well.**

Similarly, **if the rational expression H(s) has to be approximated for a specific value of the frequency** f_0, then all terms for a time-continuous circuit are evaluated in $s = j2\pi f_0$. It is now sufficient to approximate all terms contributing to the real part and all terms contributing to the imaginary part of both the numerator and the denominator up till the user-supplied maximum error ε_{max}. It can be shown quite easily that the worst-case error on the magnitude of the rational function H(s) at the frequency f_0 then is given by :

$$\frac{\Delta |H(f_0)|}{|H(f_0)|} \leq 2\ \varepsilon_{max} \tag{4.72}$$

A similar boundary can be derived for the relative error on the phase as well.

It can be concluded, that whether the approximation is performed over the whole frequency range or for a particular value of the frequency, the individual frequency coefficients are needed. In an expanded expression, these coefficients are directly available. In an expression which is nested with respect to the complex frequency, these coefficients are interwaved within the whole expression. Each term in the nested expression contributes to several coefficients, each time multiplied with a different factor. Since these factors can have different magnitudes and since it depends on the other terms in the frequency coefficient (the context) whether they are important or not, the original term in the nested expression can have an important contribution to some coefficients and not to others. As a result, this term (if it is small) may not a priori be pruned in the nested expression. This consideration is even more true if the expression is not cancellation-free, in which case an (at first sight) unimportant term can become dominant after the cancellations have been carried out. The same also holds for an expression with a nested representation of the coefficients of the complex frequency variable. If a term is important or not, depends on the factor the term is multiplied with and on the other terms (after working out the cancellations).

Example

Consider first the following expression :

$$(A + B) \ C + (C + D) \ (-A) \tag{4.73}$$

If A >> B and C >> D, one can discard B and D with respect to A and C in the sums
(A + B) and (C + D), respectively. After simplification, the false result 0 is obtained.
Instead, first the cancellations have to be carried out before performing the
approximation, resulting in the right formula :

$$B \ C - D \ A \tag{4.74}$$

A and C may not be pruned a priori although they are much smaller than B and D,
respectively. Consider now the following expression :

$$A * (C + D \ S) + B * (E + F \ S) \tag{4.75}$$

The tree of this expression is shown in Fig. 4.23. If B >> A, A still may not be
pruned. If C ≈ E, then the term B E dominates the DC coefficient. However, if D >> G
then A D dominates the first-order coefficient of S. Hence, A is important for the first-
order coefficient of S and not for the DC coefficient, and may not be pruned a priori
because it is smaller than B.   ◆

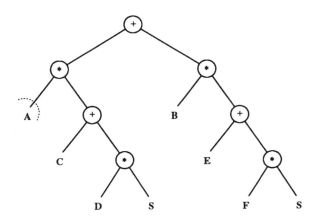

*Fig. 4.23. Tree representation of expression (4.75) with possible
pruning indicated.*

Actually, the above problems originate from the fact that during the
approximation complete terms are pruned, while the leaves in a nested expression
are symbols which contribute to several terms after multiplication with other

symbols. These symbols may not be pruned on their own, since it depends on the context and on additional cancellations whether they contribute to important terms or not.

Hence, the only possibility left for a nested expression is to extract the individual terms for all individual coefficients of the frequency variable out of the nested form. This is, however, equivalent to an expansion of the expression, thereby loosing all advantages of the nested format. This also shows why the approximation routine requires an expanded expression.

For the approximation, it is not absolutely necessary that the expanded expression is cancellation-free as well. However, it must absolutely be avoided to discard small terms which would become dominant afterwards after carrying out the cancellations. In addition, remaining cancellations confuse the interpretation of the resulting expressions and can lead to unnecessary high error estimations if the accumulated absolute error is used (as will be shown in the next subsection 4.5.2). For example, in the case of matching elements, the presence of one mismatch term clearly is more illustrative than a subtraction of two symbols, which are both much larger in magnitude than the difference between them (the mismatch term). Also, the cancellation-free form is the shortest notation for the expanded expression, and the cancellations can easily be carried out while generating the expanded form.

It can be concluded that, unless new algorithms will be found in the future to overcome the above problems without expansion, **for the given requirements within the ISAAC program an efficient approximation can only be performed on network functions in fully expanded format** :

$$H(x) = \frac{\sum_i x^i \cdot a_i(p_1, \ldots, p_m)}{\sum_i x^i \cdot b_i(p_1, \ldots, p_m)} \qquad (4.76)$$

where the polynomial coefficients $a_i(\underline{p})$ and $b_i(\underline{p})$ are in canonical SOP format and $\underline{p}$ is the vector of the symbolic circuit parameters.

4.5.2. Definitions of formula complexity and approximation error

In this subsection, the definitions of formula complexity and approximation error are provided. These definitions are important for the approximation algorithm which will be discussed in the next subsection 4.5.3. All expressions are now supposed to be in canonical SOP form.

4.5.2.1. Definition of the formula complexity

The complexity c of a canonical SOP expression is measured by the number of terms it contains. As an example. the output impedance of the bootstraped bipolar cascode output stage of Fig. 3.3 in the previous chapter is taken. The exact expression (3.6) has a complexity of 50, whereas the 10% and 25% approximations (3.7) and (3.8) have a complexity of 8 and 4, respectively.

4.5.2.2. Definition of the approximation error

There are several ways to measure the error ε introduced by approximating an expression. An appropriate definition will be derived below. The original expression is denoted as $g(p)$, the approximation as $h(p)$. All expressions are supposed to be in canonical SOP form and correspond to one of the coefficients of the complex frequency (if the approximation is carried out over the whole frequency range) or to the real or imaginary part (if the approximation is carried out for a particular value of the frequency) of either the numerator or the denominator of the network function.

A first possibility is the **worst-case error** ε_w defined as :

$$\varepsilon_w = \max_{\forall p} \left| \frac{g(p) - h(p)}{g(p)} \right| \tag{4.77}$$

where the vector of the circuit parameters p has to be varied over the total design space. This definition clearly is impractical: it takes too much CPU time and the resulting ε_w value is often much too pessimistic.

4.5.2.2.1. The nominal (or accumulated effective) error

Since the magnitude of all elements is known, corresponding to the designed values if the design has already been carried out or to estimated/default values otherwise, a **nominal error (or accumulated effective error)** ε_N can be defined in a design point p_o :

$$\varepsilon_N = \left| \frac{g(p_o) - h(p_o)}{g(p_o)} \right| \tag{4.78}$$

For formulas in SOP format, this reduces to :

$$\varepsilon_N = \frac{|\Sigma\ t_i(p_o)|}{|g(p_o)|} \tag{4.79}$$

where the $t_i(\underline{p}_o)$ are the pruned terms. This error definition gives a good estimation of the effective error in a domain around the nominal point $\underline{p}_o$.

Example

Consider again the exact expression for the DC value of the denominator of the output impedance of the bootstraped bipolar cascode stage of Fig. 3.3 :

```
g(p)  =  GO.Q1 G2 GM.Q2 Gπ.Q1 + GO.Q1 G1 G2 Gπ.Q1
       + GO.Q1 G2 Gπ.Q1 Gπ.Q2 + GO.Q1 G1 GO.Q2 Gπ.Q1
       + GO.Q1 G1 G2 GO.Q2 + GO.Q1 GO.Q2 Gπ.Q1 Gπ.Q2
       + GO.Q1 G2 GO.Q2 Gπ.Q2 + GO.Q1 G2 GO.Q2 Gπ.Q1
```
$$(4.80)$$

and the following approximation :

```
h(p)  =  GO.Q1 Gπ.Q1 G2 (GM.Q2 + G1)
```
$$(4.81)$$

In the design point :

```
p_o  =  (G1=2mS, G2=100µS, GM.Q1=GM.Q2=20mS, GO.Q1=GO.Q2=10µS,
         Gπ.Q1=Gπ.Q2=200µS)
```
$$(4.82)$$

corresponding to $I_{OUT}=0.5mA$, $\beta_F=100$, $V_{early}=50V$ and $kT/q=25mV$, the approximation $h(\underline{p})$ has a nominal error :

$$\varepsilon_N = 2.40\%$$
$$(4.83)$$

If the biasing current I_{OUT} is increased to 1mA, the nominal error in the new design point is $\varepsilon_N = 2.37\%$. If the biasing current is decreased to 0.2mA, the nominal error is now $\varepsilon_N = 2.64\%$. Hence, (4.83) is a good error estimation in a domain around $\underline{p}_o$. This is due to the tracking of all transistor parameters with the current.   ◆

The nominal error is clearly the most accurate error estimation in the nominal point and in a domain around this point (especially if all terms have the same sign). However, the nominal error is very sensitive to changes in the magnitude values in the case of large-magnitude terms with opposite sign, and it can also be much larger than 100%.

Example

A typical example is the DC value of the PSRR[+] of the CMOS two-stage Miller-compensated opamp shown in Fig. 4.24. For the following default values $\underline{p} = $ (GM=100µS, GO=1µS, ΔGM=1%.GM, ΔGO=1%.GO), the following approximated expression has a nominal error ε_N of 5.8% :

```
  2 GM.M1 GM.M2 GO.M1 GO.M4 + 2 GM.M1 GM.M2 GO.M2 GO.M4
+ ΔGM.M2BA GM.M1 GM.M3 GO.M5 - ΔGM.M1BA GM.M2 GM.M3 GO.M5
- GM.M1 GM.M3 GO.M1 GO.M5 - GM.M1 GM.M3 GO.M2 GO.M5
```
$$(4.84)$$

From this expression, it can be seen that this operational amplifier can be designed such that the contributions of the first and the second stage nominally cancel, thereby obtaining a high PSRR[+]. Hence, discarding these cancelling terms does not change the effective nominal error, but strongly reduces the insight in the cancellation mechanism and yields an underestimation of the error if parameter values differ due to process tolerances. Consider for instance the theoretical expression (A - B). Then (A - B) is as good an approximation as 0 if A has the same magnitude as B. Both formulas mathematically have the same nominal error but not the same information content. For the ISAAC program, it has been decided that the approximation algorithm has to return the formula with the largest information content for the given error, thus (A - B). The above problem is typical for large-magnitude terms with opposite sign, examples of which can be found in circuits containing parallel signal paths with similar but opposite gains. Also, if the parameter values are estimated or default values, then the actual values and the corresponding nominal error can be different. For example, if the transconductance of the output transistor M3 is doubled, then the nominal error for the approximation (4.84) becomes $\varepsilon_N = 100\%$. Moreover, the mismatch terms are statistical terms, which can be positive, zero or negative according to some probability distribution function. These mismatch terms cannot predictably cancel. The sign in the expressions only arbitrarily depends on the selected reference element. For example, if the mismatch term ΔGM.M2BA turns out to be negative but with the same value, then the nominal error again becomes $\varepsilon_N = 100\%$.   ♦

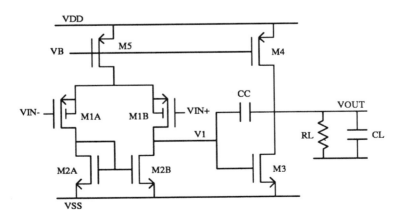

Fig. 4.24. CMOS two-stage Miller-compensated operational amplifier.

Also, if all terms in the original expression would be sorted and the smallest terms would repeatedly be removed, then the nominal error shows a zig-zag behavior with decreasing complexity if there are terms with opposite sign in the expression. This is schematically indicated in the ε-C-plot of Fig. 4.25. As a result, there is not a one-to-one correspondence between the formula complexity and the approximation error. The effective error is not a very rigid parameter to control the approximation algorithm. Several formulas of different complexity can have the same nominal error. It is the task of the approximation algorithm to return the expression with the lowest complexity for the given error.

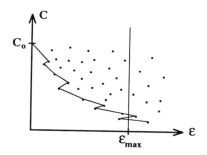

Fig. 4.25. ε-C-plot and trace if the nominal error ε_N is used as error definition.

4.5.2.2.2. The accumulated absolute error

Therefore, in [WAL_89], an alternative error definition for formulas in SOP format is presented. It is the **accumulated absolute error** ε_A, which is defined as :

$$\varepsilon_A = \frac{\Sigma \; |t_i(\underline{p}_o)|}{\Sigma \; |t_j(\underline{p}_o)|} \qquad (4.85)$$

where the $t_i(\underline{p}_o)$ are the pruned terms and the $t_j(\underline{p}_o)$ are all terms in the original expression $g(\underline{p}_o)$, all evaluated in the nominal point $\underline{p}_o$. The difference with the nominal error is that now absolute values are summed. In this way, the accumulated absolute error can largely differ from the effective error, mostly overestimating the effective error, according to the following relation :

$$\varepsilon_A \geq \varepsilon_N \frac{|g(\underline{p}_o)|}{\Sigma \; |t_j(\underline{p}_o)|} \qquad (4.86)$$

The accumulated absolute error, on the other hand, has the advantage that it is less sensitive to changes in the magnitude values for examples with large-magnitude terms of opposite sign. Consider for example again the DC value of the $PSRR^+$ of the CMOS two-stage Miller-compensated opamp of Fig. 4.24. The error for approximation (4.84) is now $\varepsilon_A = 9.7\%$. If the transconductance of the output transistor M3 is doubled, the error ε_A becomes 9.8%. If the sign of ΔGM.M2BA is reversed, then the error ε_A becomes 9.7%.

In addition, the accumulated absolute error ε_A has also an interesting property for controlling an approximation algorithm. It is always smaller than 100% and if, for a given formula, all terms are sorted and the smallest terms are repeatedly removed, then the complexity decreases monotonically with increasing error, until the maximum error is reached. This is schematically indicated in the ε-C-plot of Fig. 4.26. The trace in the ε-C-plot is now a smooth, monotonically decreasing curve. There is a one-to-one correspondence between the formula complexity and the approximation error. The accumulated absolute error clearly distinguishes formulas with the same nominal error but with a different information content. Formulas with smaller information content require a larger error ε_A. The user-supplied error now clearly corresponds to a particular approximated formula. In this sense, *if the accumulated absolute error is used, the maximum error supplied by the user is rather a parameter to control the information content of the resulting formula than a measure of the effective error.*

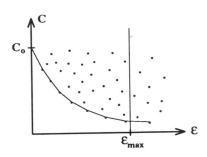

Fig. 4.26. ε-C-plot and trace if the accumulated absolute error ε_A is used as error definition.

On the other hand, this error definition requires a cancellation-free SOP form. Remaining cancelling terms would result in an unnecessary high error estimation ε_A and the difference with the nominal error would increase. For example, in the case of matching elements, an accurate error estimation is only possible by the explicit introduction of mismatch terms, since mismatch terms are much smaller than the corresponding nominal parameters. If (GM.M1B – GM.M1A) is used to represent ΔGM1BA, the error will be the sum of the magnitudes of GM.M1A and GM.M1B, which is much larger than the magnitude of ΔGM1BA.

Conclusion

It can be concluded that both error definitions have advantages and disadvantages, and thus have their applications. They are both used in the ISAAC program. The user can select which error to use during the approximation. The nominal error is the most accurate error, but it is more sensitive to parameter variations in the presence of large terms with opposite sign. It is recommended if the circuit parameters are exactly known (for example for the simulation of second-order characteristics after the circuit has already been designed nominally). The accumulated absolute error rather controls the information content of the formula and can differ from the effective error (usually by overestimating the error), but it is less sensitive to changes in parameter values. It is recommended if only estimated or default values can be used (for example for the analysis of a new circuit topology). In the next subsection, a heuristic approximation algorithm is developed, which is consistent with these error definitions.

4.5.3. The approximation algorithm

An efficient symbolic expression approximation algorithm is now presented for formulas in expanded, canonic SOP form. First, the basic approximation strategy is explained. Then, some essential extensions are discussed and the overall algorithm is presented.

4.5.3.1. The basic approximation strategy

In general two strategies are possible for the expression approximation algorithm : a global strategy and a local one.

4.5.3.1.1. Global approximation strategy

In a **global strategy**, at each step, one term is selected out of the remaining terms according to some criterion, and this term is then dropped. This is repeated until the user-supplied maximum error ε_{max} is reached :

$$\varepsilon_m \leq \varepsilon_{max} \quad \text{and} \quad C_m \text{ minimal} \tag{4.87}$$

where ε_m and C_m are the error and the complexity of the final approximated expression. The term selection criterion has to be such that always the formula with the lowest complexity for the given error is obtained. The most straightforward global technique consists of sorting all terms according to their magnitude. The smallest terms are then repeatedly removed until the accumulated error exceeds the maximum value. If the accumulated absolute error is used, this global strategy always results in the formula with the lowest complexity for the given error. Indeed, as shown in the ε-C-plot of Fig. 4.26, the bounding curve is followed, which has a smooth, monotonic behavior. If, on the other hand, the accumulated effective (or nominal) error is used and if there are terms with opposite sign in the expression, then a zig-zag curve is followed which possibly exceeds the maximum error and then tracks back below the error, as shown in the ε-C-plot of Fig. 4.25. In order to still guarantee the formula with the lowest complexity for the given error, the approximation process has to be adapted for this error definition: the approximation now has to start from the other end, the 100%-approximated expression. The largest terms are then repeatedly added until the accumulated effective error decreases below the maximum value, as shown in Fig. 4.25. The main drawback of this global approximation technique, however, is the extra CPU time consumed by the sorting routine, especially for large expressions. For example, an efficient sorting routine may consume a CPU time of the order $O(n.log(n))$ where n is the number of terms to be sorted.

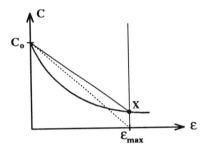

Fig. 4.27. ε-C-plot to illustrate the derivation of the local approximation criterion.

4.5.3.1.2. Local approximation strategy

In a **local strategy**, the expression is used only once. For each term it is decided whether to drop this term or not based on some local criterion. Indeed, the order in which lower-magnitude terms are dropped is of no importance as long as the expression obeys the global requirement (4.87). This fact is exploited in a local algorithm. Such an algorithm has the advantage of being much faster, but it is more difficult to predict in advance the complexity and the error of the final result [WAL_89]. The final expression must correspond to point X in the ε-C-plot of Fig. 4.27, or lie on the bounding curve close before point X. This condition is fulfilled if all terms (without sorting) are removed for which :

$$\left| \frac{dC}{d\varepsilon} \right| = \left| \frac{C_{i+1} - C_i}{\varepsilon_{i+1} - \varepsilon_i} \right| \geq \left| \frac{C_m - C_0}{\varepsilon_m - \varepsilon_0} \right| \tag{4.88}$$

where i is the index for the successive approximations, m for the final expression and 0 for the original expression, respectively. Of course, $\varepsilon_0 = 0$. The complexity of the final result C_m is not a priori known. Hence, it is replaced by the safe lower limit 0. The criterion now becomes :

$$\frac{C_i - C_{i+1}}{\varepsilon_{i+1} - \varepsilon_i} \geq \frac{C_0}{\varepsilon_m} \tag{4.89}$$

For expressions in SOP form, the complexity C_0 is equal to the number of terms in the original expression denoted by n, and the difference between two subsequent complexities is 1. The difference between two subsequent errors depends on the error definition used. For the nominal error, it is given by the magnitude of the term being discarded $|t_i(\underline{p}_0)|$ divided by the magnitude of the original expression $|g(\underline{p}_o)|$. This is also a safe upper bound in the case the accumulated absolute error is used. The final error ε_m is bounded by the user-supplied maximum value ε_{max}. Hence, the local criterion allows to drop all terms for which :

$$|t_i(\underline{p}_o)| \leq \varepsilon_{max} \frac{|g(\underline{p}_o)|}{n} \tag{4.90}$$

This means that a priori all terms $t_i(\underline{p}_o)$ can be discarded which are smaller than the fraction ε_{max} of the original expression's mean value. These terms cannot contribute to the final result for the given error ε_{max}, regardless which error definition is used. Hence, they can be dropped all together while tracing the expression only once. In addition, for formulas in SOP form, the application of this local strategy results in a point on the bounding curve. This means that it fulfils the requirement of returning an expression with the lowest complexity for the given error. However, since the above criterion (4.90) only provides a safe lower limit, the obtained expression usually has an error far below the maximum supplied value ε_{max}, thus resulting in a point on the bounding curve far before point x in Fig. 4.27.

4.5.3.1.3. Combination of both strategies

Therefore, in ISAAC, the local and the global strategies are combined, in order to speed up the overall CPU time required for the approximation. This results in the following approximation algorithm [GIE_89c, WAL_89], which is consistent with the definitions of complexity and error for canonic SOP forms given in the previous subsection. First, the expression is pre-pruned with the fast local technique (4.90). All terms smaller than the fraction ε_{max} of the original expression's mean value are dropped. The resulting expression is then fine-pruned with a global criterion until the error ε_{max} is reached. This is done by sorting the remaining terms according to their magnitude and repeatedly removing the smallest term as long as the error remains below the maximum value. The number of terms to be sorted after the local approximation is much smaller than the number of terms in the exact expression, and hence the CPU time required for the sorting and for the total approximation is strongly reduced by the local acceleration. In the ε-C-plot of Fig. 4.28, first the trace of the local strategy is followed until the point on the bounding curve is reached. From then on, the bounding curve is followed up till the maximum error is reached.

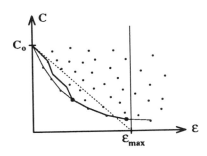

Fig. 4.28. Trace of an approximation algorithm combining a local and a global strategy.

Example

The influence of the local acceleration on the overall CPU time is now illustrated for the $PSRR^+$ of the CMOS two-stage Miller-compensated opamp of Fig. 4.24. The original expression without lumping and with mismatches in the denominator contains 258 terms in the numerator and 4292 terms in the denominator. After application of the local criterion, the number of terms is reduced to 72 and 587 for the numerator and the denominator, respectively. The final result for an accumulated absolute error of 25% contains 20 terms in the numerator and 40 terms in the denominator (for all powers of s). The CPU times for this example on a TI EXPLORER I (~1 Mips) with and without local acceleration are summarized in the following table :

	without acceleration	with acceleration
numerator denominator	1.9s 72.6s	1.1s 24.9s

Clearly, the local acceleration offers a substantial CPU time reduction.   ◆

4.5.3.2. Extensions to the basic approximation strategy

The above algorithm, as described in [WAL_89], still has two drawbacks. First of all, this algorithm in combination with the accumulated absolute error can still result in large effective error values (even for a small maximum error value) and can return very confusing results, in the presence of large-magnitude terms with opposite sign. This is illustrated in the following example.

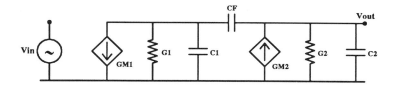

Fig. 4.29. Example AC circuit to illustrate some difficulties for an approximation algorithm.

Example

Consider the equivalent AC circuit of Fig. 4.29. For the following magnitude values (GM1=2mS, R1=20kΩ, C1=5pF, CF=20pF, GM2=50µS, R2=1MΩ, C2=5pF), the above algorithm returns the following expression for an accumulated absolute error of 10% :

$$\frac{-\ \text{GM1}\ (\text{GM2}\ +\ \text{S}\ \text{CF})}{\text{G1}\ \text{G2}\ +\ \text{S}\ (-\ \text{GM2}\ \text{CF}\ +\ \text{G1}\ \text{CF}\ +\ \text{G1}\ \text{C2})\ +\ \text{S}^2\ (\text{C1}\ \text{CF}\ +\ \text{C2}\ \text{CF}\ +\ \text{C1}\ \text{C2})} \tag{4.91}$$

If the error is increased to 13%, the following expression is returned :

$$\frac{-\ \text{GM1}\ (\text{GM2}\ +\ \text{S}\ \text{CF})}{\text{G1}\ \text{G2}\ +\ \text{S}\ \text{CF}\ (-\ \text{GM2}\ +\ \text{G1})\ +\ \text{S}^2\ \text{CF}\ (\text{C1}\ +\ \text{C2})} \tag{4.92}$$

Both expressions have approximately the same complexity but a totally different behavior as shown in Fig. 4.30, more than the 3% increase in error can justify. The reason is found in the coefficient of s in the denominator. The terms (- GM2 CF) and (+ G1 CF) have exactly the same large magnitude, but an opposite sign. Hence, they numerically cancel and the circuit's behavior is determined by the other, smaller term G1 C2. However, this term is dropped in the 13%-approximated expression, because the accumulated absolute error adds absolute values and |G1 C2| is much smaller than |GM2 CF| + |G1 CF|. For this example, a 3% increase in error drastically changes the circuit's behavior.   ◆

This problem can be solved with a simple extension to the above algorithm. It can intuitively be seen that, if the maximum allowed error is ε_{max}, all terms with a magnitude larger than the fraction ε_{max} of the expression's effective value $|g(\underline{p}_o)|$ may not be dropped. This condition puts an upper limit on the magnitude of terms which may be pruned, in the same way as the local criterion puts an upper limit on the magnitude of terms which may be pruned a priori. Hence, the process of repeatedly removing the smallest term is stopped when the maximum error value is reached or when the magnitude of the term becomes larger than the following limit :

$$|t_i(\underline{p}_o)| \geq \varepsilon_{max} |g(\underline{p}_0)| \tag{4.93}$$

In the presence of large-magnitude terms with opposite sign, such as in the above example, the expression's magnitude $|g(\underline{p}_o)|$ is small and hence the limit (4.93) is adapted appropriately. Note that this upper limit also holds in the case the nominal error is used.

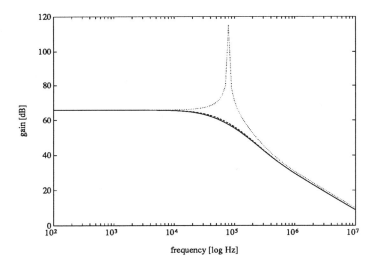

Fig. 4.30. Plot of the gain of the circuit of Fig. 4.29: exact expression (straight line), 10% approximated expression (dashed line) and 13% approximated expression (dotted line).

The second remark on the above approximation algorithm comes from the requirement of providing the expression with the largest information content. If a practical expression is observed, then the magnitudes of the terms typically appear in clusters. The magnitudes of all terms for the DC value of the PSRR[+] of the CMOS two-stage Miller-compensated opamp of Fig. 4.24 are shown in Fig. 4.31. Many terms have the same or nearly the same magnitude (sometimes differing by a factor

the $(\alpha_1{}^p * \alpha_2{}^q)$-order approximation of the network function. To each order corresponds a certain weight, depending on the values of α_1 and α_2 and their exponents. The NETFORM program [SMI_78b] then returns the nonzero approximation with the lowest order in α_1 and α_2. Therefore, it first calculates the $(\alpha_1{}^0 * \alpha_2{}^0)$-order approximation of the requested network function. If this is zero, it calculates the $(\alpha_1{}^1 * \alpha_2{}^0)$-order and/or the $(\alpha_1{}^0 * \alpha_2{}^1)$-order approximation. If these are zero as well, it calculates the $(\alpha_1{}^1 * \alpha_2{}^1)$-order approximation. Each of these approximations can be generated quite easily during the development of the determinant. The efficiency is improved by the early truncation of terms which cannot contribute to the order being developed for. Cofactors of truncated elements are not calculated. Also, the truncation is applied recursively to all minors. This technique will now be illustrated in the following example.

Example

Consider the following *3x3*-matrix with the two small-magnitude symbols α_1 and α_2 :

$$
\begin{bmatrix}
1 & \alpha_1 & 0 \\
0 & 1 & \alpha_2 \\
A & B & C
\end{bmatrix}
\tag{4.95}
$$

The determinant of this matrix within order $(\alpha_1{}^0 * \alpha_2{}^0)$ is given by :

$$
1 * \begin{vmatrix} 1 & \alpha_2 \\ B & C \end{vmatrix} - \alpha_1 * \ldots
$$

$$
= 1 (1 * C - \alpha_2 * \ldots) = C
\tag{4.96}
$$

All terms containing α_1 or α_2 are truncated during the development. Also, during the recursive minor expansion process, preference is given to developing the minors along rows containing small parameters. On the other hand, if the same determinant is calculated within order $(\alpha_1{}^1 * \alpha_2{}^0)$ or $(\alpha_1{}^0 * \alpha_2{}^1)$ (suppose both α_1 and α_2 have the same magnitude), then the following result is obtained :

$$
1 * \begin{vmatrix} 1 & \alpha_2 \\ B & C \end{vmatrix} - \alpha_1 * \begin{vmatrix} 0 & \alpha_2 \\ A & C \end{vmatrix}
$$

$$
= 1 (1 * \ldots - \alpha_2 * B) - \alpha_1 (- \alpha_2 \ldots)
$$

$$
= - \alpha_2 * B
\tag{4.97}
$$

♦

This power series truncation, however, has some serious drawbacks.

- First of all, the expression is always truncated up to a certain order, no matter how many terms of that order are involved and no matter what the values of the truncated terms are [SMI_78b]. Hence, the order is not directly related to the error introduced by the approximation and the method does not guarantee at all that the most important terms are returned to the user. For example, if an expression contains one small-magnitude symbol α, it can be written as :

$$A + B * \alpha \qquad\qquad\qquad (4.98)$$

The above truncation algorithm then returns A (if it is nonzero), regardless of the value of B. However, in reality, B $*$ α can be much larger than A. See for example the mismatch terms which are as important as the other terms in (4.84).

- Secondly, this truncation algorithm only develops a discrete number of orders. It does not allow to continuously change the approximation error, and it certainly does not always return the most dominant contributions for the given error. For example, the most important terms (if the magnitudes of all elements are considered) could be some of the terms of A and some of the terms of B $*$ α. Moreover, the method is difficult to extend to expressions containing the frequency as a symbol. The truncation can only be carried out for some specific value of the frequency and not over the whole frequency range. Clearly, the algorithm presented in [SMI_78b] is not suited for the needs of analog designers, whereas these requirements are all fulfilled in the ISAAC program.

- Thirdly, for the NETFORM program, the user himself has to prepare the circuit equations into an appropriate form, such that large values are replaced by their reciprocal small value, and that values close to a specific value, say 1, are represented by this value minus a small difference, for example $1-\alpha$, and that these small parameters coincide with matrix entries [SMI_78b].

It can be concluded that the truncation algorithm presented in [SMI_78b] is fast, since the truncation is performed during the development of the determinant. However, the results can be unreliable and are definitely not adapted to the needs of analog designers.

4.5.4.2. SYNAP

The SYNAP program [SED_88] also contains some features for the simplification of large, exact expressions. The techniques used for this simplification are the following :
- exploiting differential-pair and current-mirror information
- dropping parasitic capacitors

- pruning insignificant subexpressions based on relative sizes of circuit parameters.

These techniques are now discussed shortly and compared to the features provided in the ISAAC program.

The exploitation of differential-pair information is similar to the matching concept in the ISAAC program. The small-signal elements of both transistors are represented by the same symbols. The exploitation of current-mirror information is different. The small-signal elements of the output transistor are then represented by B times the small-signal elements of the input transistor, where B represents the current-mirror ratio. This greatly simplifies the calculation and eases the interpretation of the results. However, it leads to very large errors, much larger than the supplied maximum approximation error. For example, for the CMOS OTA of Fig. 4.8 and with the exploitation of current-mirror information, [SED_88] reports that the unpruned (unapproximated) equation varies from SPICE results by 33% at lower frequencies, since the output conductances in reality do not follow the current-mirror ratio. Hence, this technique is not used in ISAAC.

The dropping of all parasitic capacitors comes to ignoring all transistor small-signal capacitances. The only capacitors incorporated into the analysis are then external capacitors, such as the load capacitor. The same results can be obtained with ISAAC by selecting a DC model for the transistors. However, it is a rather crude all-or-nothing approximation, which seriously affects all zeros and poles in the system (especially at higher frequencies). Besides, for a MOS transistor in saturation, the gate-source capacitor is much larger than the gate-drain capacitor. This relative size information could better be exploited by dropping the gate-drain capacitors only.

Finally, the expressions are also pruned based on the relative sizes of the circuit parameters. As in ISAAC, each coefficient of the complex frequency s in the numerator and denominator is pruned individually such that it changes by at most the user-specified percentage. However, **two pruning methods are available in SYNAP: tree pruning and flat pruning.**
- tree pruning is the fastest method. It acts directly on the expression's tree. But it has the drawbacks mentioned before in subsection 4.5.1. Indeed, it has been observed [SED_89] that SYNAP sometimes prunes dominant terms, while less important terms are still present. This largely confuses the designer who uses the program. The error is also more difficult to control.
- for flat pruning, on the other hand, the expression is first expanded into a sum-of-products form and the smallest terms are then dropped. This is similar to the global approximation technique in ISAAC. It requires more CPU time than tree pruning, but it always returns the dominant terms [SED_89] and the error can be controlled much better.

However, either which pruning method is used in SYNAP, the exact transfer function, which is in nested form, must first be converted into a form suitable for pruning. This step takes a considerable amount of CPU time [SED_88] and is not needed in the ISAAC program since the equation solution algorithm immediately generates the expressions in the right format. Moreover, detailed experiences [SED_89] have shown that SYNAP definitely requires more CPU time than ISAAC for flat pruning the same examples. Both programs are compared in the following table. The CPU times are given for the $PSRR^+$ of the CMOS two-stage Miller-compensated opamp of Fig. 4.24 with all internal capacitances included.

	type of pruning	CPU time	quality of result
ISAAC	flat pruning	2.51 s	+
SYNAP	tree pruning	2 s	-
	flat pruning	100 s	+

The results of the tree pruning in SYNAP are unreliable. Moreover, this tree pruning requires approximately the same CPU time as the flat pruning in ISAAC. This clearly indicates the efficiency of the approximation algorithm in the ISAAC program.

4.5.4.3. FORMULA

Finally, in FORMULA [RAA_88], an alternative approximation technique is applied. First, the largest term in each coefficient of the complex frequency variable s is traced. All terms smaller than a user-supplied percentage of this largest term are then dropped. This technique is applied in two ways: after the whole expression has been calculated (similar to flat pruning), or during the calculation of the expression (similar to tree pruning). The latter method introduces an additional error. For the experiments mentioned in [RAA_88], this additional error is below 5% of the frequency coefficient's total value for a user-supplied error of 1%. However, this still can lead to erroneous results, since it is likely possible that dominant terms are pruned during the calculation of the determinant if there are any cancelling terms in the expression. Besides, these cancellations are inherent to the nodal admittance formulation used in [RAA_88]. Moreover, with the error merely indicating the fraction of the largest term below which all terms are dropped, it is difficult to know in advance the overall error introduced by this approximation. The user clearly has less control over the error for the different frequency coefficients and thus on the overall approximation error.

Conclusion

These comparisons to other approximation approaches presented in the literature clearly show the efficiency of the approximation algorithm used in ISAAC. In addition, it is also shown that this algorithm is most suited to the needs of analog designers, since it allows to continuously decrease or increase the error value and since it always returns the most dominant terms in the expression.

4.5.5. Conclusion

In this subsection, an algorithm has been presented for the heuristic approximation of symbolic expressions in expanded format. The expanded format is needed to carry out the approximation over the whole frequency range or for a particular center frequency value, while always returning the most dominant terms in the result. Several definitions have been given for the error introduced by the approximation. The nominal error (or accumulated effective error) is the most accurate estimation of the approximation error. The accumulated absolute error, on the other hand, is less sensitive to the actual magnitude values used for the approximation and it is thus preferred if the exact values are not yet known. The accumulated absolute error rather controls the information content of the expression. The approximation algorithm itself combines a local and a global approximation strategy, with some extensions to cope with large-magnitude terms of opposite sign. The algorithm is then compared to the few other approaches presented in the literature. These comparisons clearly indicate the efficiency of the approximation algorithm and they also show that the approximation in ISAAC highly suits the needs of analog designers.

4.6. Performance of the ISAAC program

Finally, in this section, the overall performance of the ISAAC program is summarized and compared to similar programs from the literature. In the first subsection 4.6.1, the performance of the ISAAC program is evaluated and the influence of matching and lumping on the CPU time are shown. In the second subsection 4.6.2, the ISAAC program is then compared to other programs from the literature with respect to methods used, functionality and efficiency. It will be shown that ISAAC is the most functional symbolic analysis tool for analog designers and that it presently outperforms all other similar tools.

4.6.1. Performance evaluation of the ISAAC program

In this section the performance of the ISAAC program is measured for several examples with increasing size. All CPU times are given for a TI EXPLORER II Plus LISP-station (~10 Mips, 24 Mbyte physical memory, 128 Mbyte virtual memory). The time for the set-up of the equations is negligible for all examples. Most of the CPU time is consumed by the equation solution routine and by the approximation routine. Therefore, these two times are given below for all examples, together with the CMNA matrix size, the number of symbolic elements and the number of symbolic terms generated. Note also that all data are split up in data for the numerator (the number before the backslash) and data for the denominator (the number after the backslash) of each requested transfer function.

The increase in the CPU time required by the equation solution algorithm as a function of the circuit size has already been investigated in 4.4 for an extremely sparse and an extremely dense example. There, it has been shown that the solution algorithm requires a CPU time which increases linearly with the number of terms in the resulting expression. The increase of the number of terms with the circuit size depends on the actual circuit topology and the number of symbolic circuit elements, but usually is exponential. The CPU time required by the approximation routine also increases about linearly with the number of terms in the expression and depends on the value of the approximation error.

The first series of examples considered here to measure the overall performance of the ISAAC program are active RC filters and switched-capacitor filters. The filters being analyzed are three active RC filters, the Vogel filter of Fig. 4.12 [LIB_88] (filter1), the lowpass filter of Fig. 4.7 [STA_86] (filter2) and the bandpass filter of Fig. 4.14 [STA_86] (filter3), and two switched-capacitor filters, the Fleischer-Laker biquad of Fig. 4.10 [FLE_79] (filter4) and a fifth-order switched-capacitor filter [WAL_91] (filter5). For all these filters, the transfer function is analyzed without lumping and matching and with ideal operational amplifiers. For filters of course only the exact filter function is important without any approximation. The results are summarized in the following table.

filter	solution time	matrix size	nr elements	nr terms
filter1	0.012/0.005	3x3	10	5/3
filter2	0.021/0.008	5x5	13	11/6
filter3	0.22/133	18x18	44	108/6840
filter4	0.026/0.076	4x4	12	11/8
filter5	0.93/12.02	10x10	23	36/357

The number of terms and the CPU time strongly increase with the size of the filter circuit. However, the CPU times are still acceptable for practical filter configurations. For example, the fully symbolic analysis of the bandpass filter of Fig. 4.14 requires only 133 s for a circuit with 33 nodes and 44 symbolic elements, and the symbolic analysis of a fifth-order switched-capacitor filter requires only 13 s. The situation however is worse for semiconductor circuits, which have a small-signal expansion and thus a large number of symbolic elements for a relatively small number of nodes, since for instance every MOS transistor by default is expanded in (at most) 7 symbolic elements. As a result, the maximum number of nodes which can be analyzed in practice is smaller for semiconductor circuits than for filter circuits. This is now investigated in the following examples.

The second series of examples considered to measure the overall performance of the ISAAC program are CMOS operational amplifiers with increasing number of transistors. The amplifiers being analyzed are a simple OTA (opamp1), a two-stage Miller-compensated opamp with simple OTA input stage (opamp2, Fig. 4.24), a symmetrical OTA (opamp3, Fig. 4.8), a fully differential Castello opamp [WAN_86] (opamp4), a folded-cascode OTA (opamp5), and a two-stage Miller-compensated opamp with symmetrical OTA input stage (opamp6). For all opamps, the differential gain is analyzed with all parasitic capacitances included, with lumping of like parallel elements and with exploitation of matching information (but without including explicit mismatch terms). Note that the symmetry of the differential Castello opamp (opamp4) is exploited to reduce the CPU time. The exact expressions are then approximated with a 25% error. The results are summarized in the following table.

opamp	solution time	pruning time	matrix size	nr elements	nr terms
opamp1	0.30/0.29	0.04/0.08	3x3	26	60/110
opamp2	0.75/1.00	0.10/0.36	4x4	37	120/375
opamp3	4.99/13.37	0.57/2.22	5x5	44	648/2202
opamp4	2.6/5.5	0.7/1.6	5x5	39	640/1680
opamp5	55.5/106.1	5.2/24.6	6x6	51	5292/12523
opamp6	16.7/51.3	1.3/17.9	6x6	55	1296/8403

The CPU times for the above operational amplifiers are logarithmically plotted in Fig. 4.34 as a function of the number of transistors. The number of terms and hence the CPU time rapidly increase for larger circuits. The fully symbolic expression for the differential gain of the CMOS two-stage Miller-compensated opamp contains 495 terms (both numerator and denominator) for a circuit with 4 internal nodes and 7 transistors (37 symbolic elements) and is generated in 1.75 s. The expression for the symmetrical OTA with 5 internal nodes and 9 transistors (44 symbolic elements) already contains 2850 terms and is generated in 18.4 s. The number of terms and the analysis time clearly increase exponentially with the circuit size. This is an observation which generally holds for the generation of fully symbolic expressions in expanded format, especially for linearized (small-signal) circuits. Note also that the analysis time not only depends on the number of transistors, but also on the circuit topology (the interconnection of these transistors). For example, although opamp6 has one transistor more than opamp5, it is analyzed in a shorter time.

The number of terms and the CPU time can be decreased by lumping like parallel elements and by exploiting matching between elemens. This **influence of lumping and matching on the CPU time and storage requirements** is illustrated for the differential gain of the CMOS two-stage Miller-compensated opamp of Fig. 4.24. This gain has been analyzed with and without the exploitation of matching information, each time with and without lumping of like parallel elements. The CPU time (on a TI EXPLORER II Plus (~10 Mips)) and the total number of terms in the expression (both numerator and denominator summed) are summarized in the following table for the four analyses :

	without lumping	with lumping
without matching	15.7 s/7533 terms	3.1 s/982 terms
with matching	7.4 s/3469 terms	1.7 s/495 terms

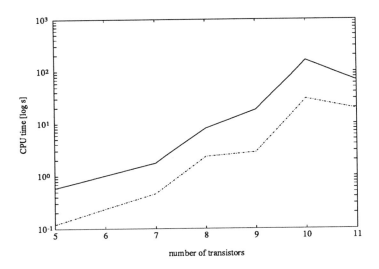

Fig. 4.34. CPU time for the equation solution (straight line) and the expression approximation (dashdotted line) for CMOS operational amplifiers as a function of the number of transistors.

Clearly, both transformations, lumping and matching, strongly reduce the CPU time and the complexity of the result. The exploitation of matching information approximately halfs the CPU time and the number of terms. The lumping of like parallel elements reduces the CPU time by a factor of 5, and the number of terms by a factor of 7-8, respectively. Still, the remaining number of terms, 495, is quite large for a circuit with only 4 internal nodes and 7 transistors (37 symbolic elements). Lumping and matching reduce the number of terms and the CPU time, but the exponential increase with the circuit size remains.

As a result, it can be concluded that symbolic simulation has a limitation with respect to the maximum circuit size which can be analyzed in practice. The maximum circuit size which ISAAC can practically handle depends on the number of terms generated in the analysis, which depends on the number of nodes and the number of symbolic circuit elements. This limitation of course depends on the processor speed of the computer used and the available amount of memory (storage). For the present TI EXPLORER II Plus configuration (~10 Mips, 24 Mbyte physical memory, 128 Mbyte virtual memory) this **limit is somewhat around 40 nodes for filter circuits and around 15 transistors for semiconductor circuits for a fully symbolic analysis in expanded format with all parasitic capacitances included.** However, as with all programs, one must learn to use the program appropriately for the targets it has been developed for. A symbolic simulator will never replace a numerical simulator. **A symbolic simulator and a numerical simulator are complementary tools which provide a different perspective on the same circuit to the designer. Both tools also will never exclude the reasoning and the insight of the human designer. Only, symbolic simulation, if applied appropriately, is more suited for helping the designer in gaining insight into a circuit's behavior.** For large circuits, the symbolic expressions are cumbersome and hard to interpret anyway. For example, most high-frequency terms in large, fully symbolic expressions are of no practical use and only complicate the interpretation of the results. Hence, the analysis can highly be speeded up and the interpretability can highly be improved by modifying the small-signal models of the transistors and excluding small parasitic capacitances from the analysis. The analysis time for the folded-cascode OTA (opamp5) without parasitic capacitances is 15 s. In addition, 15 to 20 transistors typically is the maximum circuit complexity handled by analog expert designers as one (sub)block. For larger circuits more insight can be gained by hierarchically decomposing the circuit into smaller subcircuits, analyzing each subcircuit separately with the techniques presented in this chapter and then analyzing an equivalent schematic for the original circuit, thereby modeling each subcircuit. The final circuit size aimed at with the symbolic simulator is of the order of practical building blocks, such as amplifiers, buffers, leapfrog filters, etc. ISAAC, however, has already well proven itself by returning correct symbolic expressions, which often differed from an expert designer's expectations, even for small circuits. In addition, the program can be used for larger circuits as well if it will be combined with a hierarchical decomposition method. Finally, for the generation of analytic circuit models, the expressions can be much larger since they are just used for numerical evaluation during the circuit sizing (optimization). For this evaluation, the expressions even do not necessarily have to be expanded, but can be kept in exact, unexpanded format as well. Most of the parameters, however, have a really negligible influence on the result. The approximation of the expressions can strongly reduce the numerical evaluation time during the optimization and has to be carried out only once for each analytic model. This trade-off will be investigated in more detail in chapter 6. **The use of formulas in nested format will overcome the exponential increase of the CPU time with the circuit size. Up till now, however, no adequate approximation algorithm has been found yet for the approximation of nested formulas.** This nesting of formulas and other techniques to extend the ISAAC program to larger circuits will be described in more detail in the next chapter. It is however important to mention that ISAAC has already proven itself for many smaller circuits as well. It is extensively being used for the analysis of operational amplifiers and switched-capacitor filter, but it can be applied to other classes of analog circuits as well.

In this subsection, the performance of the ISAAC program has been evaluated. This performance and the functionality of the program are now compared to other symbolic analysis programs in the next subsection.

4.6.2. Comparison to other symbolic analysis programs

The symbolic analysis techniques presented in section 4.2 have resulted in several symbolic analysis programs in the past. Most of these programs, however, generate exact expressions only and are not tuned to the needs of analog designers. The *ISAAC* program [GIE_89c, SAN_89, WAL_89] on the other hand is dedicated to analog design, provides a built-in expression approximation and combines the analysis of time-continuous and time-discrete circuits in one and the same program. In this subsection, the most important symbolic analysis programs presented in the literature are described shortly. They are then compared to the ISAAC program with respect to functionality, methods used and efficiency.

The *NETFORM* program [SMI_81a, SMI_81b] performs a symbolic analysis based on a modified tableau formulation, in which the circuit equations are made as sparse as possible by including extra variables. It is implemented in the general symbolic mathematics package REDUCE and can generate both expanded and nested expressions with a sparse recursive minor expansion algorithm. NETFORM has a primitive form of approximation which only allows for discrete orders to be truncated. The program can analyze linear time-continuous circuits only. No small-signal expansion or matching are provided.

The *SYNAP* program [SED_88] is also directed to analog integrated circuits. It performs an approximated symbolic DC analysis and generates symbolic network functions for both linear and linearized time-continuous circuits. The AC analyzes are based on the MNA method and a general symbolic algebra package is used for the determinant calculation. This package treats the formulas hierarchically, but they can be expanded later on for the approximation. SYNAP provides some simplification features such as the inclusion of matching between elements, and it also provides two types of expression approximation, the fast but unreliable tree pruning and the slower flat pruning. The SYNAP program is still under development.

The *SAPEC* program [LIB_88] generates symbolic expressions for linear circuits (filters) in the s-domain only. It has no small-signal expansion and no approximation. It is based on the MNA method and uses the nonzero-permutation generation method for the determinant calculation. It is however a small program on a personal computer with a nice (also graphical) interface. SAPEC can also generate symbolic sensitivity functions and testability figures for the linear circuits [LIB_89].

In *CASNA* [DEL_89], the circuit equation formulation of Sannuti and Puri [SAN_80] is used which results in a matrix with symbolic elements on the diagonal only. The determinant of this matrix is calculated by means of the determinant definition. CASNA is an experimental program with a limited set of primitives. It can only handle time-continuous linear circuits, has no small-signal expansion and no approximation.

FORMULA [RAA_88] is another experimental symbolic analysis program for time-continuous analog circuits. It has a small-signal linearization, but only knows a limited set of primitives. The program uses the nodal analysis matrix which is then reduced to an equivalent two-port matrix. A limited form of approximation is provided.

All other symbolic analysis programs presented in the past are more originating from circuit theory research, and are not useful for analog designers as they only provide an exact symbolic analysis of linear circuits, without any small-signal linearization nor any approximation. For example, *TOPSEN* [ADB_80] uses some form of tree enumeration, but it produces erroneous results for some examples as verified both by ISAAC [GIE_89c] and CASNA [DEL_89]. [STA_86] combines the signal-flow-graph method with hierarchical decomposition and a nested formula representation, which extends the capabilities of the program to larger circuits. This can be a useful technique on its own, but for analog integrated circuits the approximation of the expressions is really necessary. The approximation, which makes the program useful for analog designers, is precisely one of the novel features introduced by ISAAC [GIE_89c].

With respect to the symbolic analysis of switched-capacitor networks, several approaches have been presented as well [BON_80, CHE_87, GIE_89c, JOH_84, KON_88, MOS_84,86,89, PUN_85, TAN_82, WAL_89,91, XIA_85]. One of the most viable programs is *SCYMBAL* [KON_88] which uses a signal-flow-graph

	ISAAC	SYNAP	SAPEC	CASNA	FORMULA	NETFORM
analysis types	s-domain & z-domain	DC & s-domain	s-domain	s-domain	s-domain	s-domain
primitives	complete set	limited set	complete set	limited set	limited set	complete set
small-signal linearization	yes	yes	no	no	yes	no
nonlinear analysis	yes	no	no	no	no	no
implementation language	LISP	LISP + MACSYMA	LISP	PASCAL	LISP	REDUCE
expressions format	expanded	expanded/nested	expanded	expanded	expanded	expanded/nested
equation formulation	CMNA	MNA	MNA	Sannuti-Puri formulation	nodal analysis	sparse tableau
equation solution algorithm	recursive minor expansion	determinant calculation	permutation generation	determinant definition	matrix reduction	recursive minor expansion
approximation	yes	yes	no	no	limited	primitive

Table 4.1. Comparison of symbolic analysis programs: ISAAC, SYNAP, SAPEC, CASNA, FORMULA and NETFORM.

method based on the MNA formulation of the circuit equations. The program can analyze any switched-capacitor circuit and can symbolically represent the frequency variable z, the capacitors, any opamp gain and the timing functions for the switches. However, similar capabilities are provided in the ISAAC program [GIE_89c,

WAL_89], and, up till now, ISAAC is the only program which allows the symbolic analysis of both time-continuous and time-discrete circuits in one and the same program.

The symbolic analysis programs discussed above are now compared in more detail in table 4.1 with respect to functionality and methods used. Clearly, the ISAAC program has the largest functionility for analog designers. ISAAC [GIE_89c], SYNAP [SED_88] and FORMULA [RAA_88] are the only programs which have a small-signal linearization and an expression approximation. SYNAP also offers a symbolic DC analysis [SED_88], but ISAAC additionally provides the analysis of switched-capacitor circuits [WAL_89] and a symbolic harmonic distortion analysis [WAM_90].

Finally, the **overall efficiency of the ISAAC program is compared to the other programs**. This is however a difficult task since most CPU time values in the literature are published for :
- different computers (processor and memory configuration)
- different examples.

It would be interesting to propose a set of standard examples as benchmark for any symbolic analysis program in the future. In the mean time, ISAAC can only be compared experimentally to the other programs for some individual examples. The CPU times for ISAAC are given for the TI EXPLORER II Plus configuration.

The Tow-Thomas filter [TSA_77] is analyzed by several programs: ISAAC requires 6 ms, SAPEC [LIB_89] 6.86 s on an IBM PS/2 Model 30, [SIN_77] more than 0.49 s on an IBM 370/158 and SNAPEST [TSA_77] 0.282 s on a CDC CYBER 74. Another example analyzed by SAPEC [LIB_88] is the Vogel filter of Fig. 4.12: it requires 7.74 s for SAPEC on a PC against 17 ms for ISAAC on a LISP-station. The active RC filter of Fig. 4.7 is analyzed by ISAAC in 29 ms, whereas it takes 83 ms in [STA_86] on a DTS/8 GCOS. NETFORM analyzes the gyrator circuit presented in [SMI_79] in 81 ms on a DEC PDP-10 against 24 ms for ISAAC. If we compare ISAAC to CASNA, then the largest example presented in [DEL_89] is a passive RC filter which takes 16 s for CASNA on a VAX 780 against 0.52 s for ISAAC. These figures experimentally show the efficiency of the ISAAC program. They also indicate that the performance of ISAAC (which is determinant-based) is at least comparable to that of topological methods. This conclusion is not surprising since there must always be some equivalence relation between a matrix formulation and a graph formulation of the same problem.

Another often used example is the resistor ladder network of Fig. 4.16. For example, a ladder with ten sections is analyzed by CASNA [DEL_89] in 1330 s on a VAX 780, whereas ISAAC requires 86 s on the TI EXPLORER II Plus. However, for this example, ISAAC cannot beat the programs which generate expressions in nested format: [SMI_79] analyzes 12 sections in about 2.5 s on a DEC PDP-10, whereas [STA_86] even analyzes 90 sections in about 1.5 s on a CDC Cyber 73.

Another large example is the bandpass filter of Fig. 4.14. This filter requires 133 s for ISAAC in expanded format. The same filter is analyzed in nested format by [STA_86] and [HAS_89a]. No CPU times have been published, however, and the result presented in [STA_86] is even wrong. In addition, these programs do not allow for any approximation. With respect to switched-capacitor circuits, the biquad of Fig. 4.10 only takes 0.102 s, whereas a fifth-order filter is analyzed by ISAAC in 13 s.

As already discussed above, three symbolic analysis tools provide capabilities (small-signal linearization, approximation) for analyzing semiconductor circuits: ISAAC [GIE_89c], SYNAP [SED_88] and FORMULA [RAA_88]. FORMULA analyzes and approximates the CMOS two-stage Miller-compensated opamp of Fig. 4.24 in 6.4 s on a VAX computer. The same opamp is analyzed and approximated by ISAAC in 2.2 s. The operational transconductance amplifier (OTA) of Fig. 4.8 is analyzed and approximated by the three programs. ISAAC requires 21 s for a fully symbolic analysis with all parasitic capacitors included. This is below the 62.4 s needed by FORMULA [RAA_88] on a VAX computer, and far below the 32 minutes required by SYNAP [SED_88] on a SYMBOLICS machine. Besides, although the CPU times of FORMULA are quite comparable to those of ISAAC, the approximation, the available primitives and the analysis capabilities are very limited in FORMULA.

The above comparisons indicate that **ISAAC presently outperforms all other symbolic analysis tools, which generate expressions in expanded format**. This expansion is required for the approximation, which is an indispensable feature for the symbolic analysis of analog integrated circuits.

4.6.3. Conclusion

In this section, the overall performance of the symbolic simulator ISAAC [GIE_89c, SAN_89] has been evaluated. It has been shown that ISAAC is the most functional symbolic analysis tool for analog designers, since it provides a small-signal linearization, a wide range of analysis types and a built-in expression approximation. ISAAC is also the only program which combines the symbolic analysis of time-continuous and time-discrete circuits in one and the same program, and which provides a symbolic distortion analysis. This distortion analysis will be discussed in the next chapter. It has also been shown experimentally that ISAAC's efficiency is better than or at least comparable to other approaches (as far as expanded expressions is concerned). Typical CPU times are 1.75 and 13 seconds for the voltage gain of a CMOS two-stage opamp and of a fifth-order switched-capacitor ladder filter, respectively. Summarizing, it can be concluded that ISAAC presently outperforms all other tools for the symbolic analysis of analog integrated circuits.

4.7. Conclusions

In this chapter, the algorithmic aspects of symbolic simulation in general and of the symbolic simulator ISAAC [GIE_89c, SAN_89, WAL_89] in particular have been discussed. ISAAC generates symbolic expressions for the AC characteristics of analog integrated circuits. These expressions can then be simplified with a heuristic criterion based on the relative magnitudes of the elements. The expressions are thus generated by consecutively setting up the circuit equations, by symbolically solving these equations and by approximating the resulting expressions.

First, an historical overview of symbolic analysis has been given. The different symbolic analysis techniques presented in the literature have been discussed and compared with respect to their usefulness and efficiency. For the ISAAC program a determinant calculation method has been adopted and it has been shown in this chapter that determinant-based methods can be as efficient as topological (graph-based) methods.

Then, the set-up of the linear circuit equations has been discussed. For the ISAAC program, a compacted modified nodal analysis (CMNA) formulation is used, which yields compact and still sparse matrices. As compared to the tableau and the MNA methods, much smaller matrix sizes are obtained with the CMNA method due to the presented row and column compactions, which allow the elimination of several variables and circuit equations. As compared to the MNA method, symbolic term cancellations are reduced or even taken away. The CMNA method thus offers a fair compromise between matrix compactness and term cancellations. In addition, the method is generally applicable, also to switched-capacitor circuits, and the polynomial format of all matrix entries is maintained by the compactions. All these features allow for an efficient solution of the symbolic network equations.

For the symbolic solution of the CMNA equations, a dedicated and fast determinant expansion algorithm has been developed. It exploits the mathematical structure of the sparse CMNA equations and generates the expressions in expanded canonic SOP format. This format is imposed by the expression approximation routine. The equation solution algorithm itself is based on a recursive double sparse Laplace expansion of the determinant with storage of minors. This algorithm has been shown to outperform all other determinant algorithms for the given requirements.

The equation solution routine returns exact, fully expanded expressions. For practical analog circuits, these expression are usually lengthy and difficult to interpret. Therefore, a heuristic approximation has been included in the ISAAC program which approximates (or prunes) the calculated symbolic expressions up to a user-defined error based on the relative magnitudes of the elements and which returns the dominant terms in the result only. In a short introduction, symbolic expression approximation has been depicted as a trade-off between formula complexity and formula error. The approximation routine within the ISAAC program has to return the expression with the lowest complexity but with the largest information content for the given error. In order to carry out the approximation over the whole frequency range or for a particular center frequency while always guaranteeing to return the most dominant terms in the result, an expanded format is required for the expressions. For measuring the error introduced by the approximation, several definitions have been given and compared. The nominal error (or accumulated effective error) is the most accurate estimation of the approximation error, but the accumulated absolute error is less sensitive to the actual magnitude values and is thus preferred if the exact values are not yet known. The accumulated absolute error rather controls the information content of the expression. The approximation algorithm itself follows a global approximation strategy, which is speeded up by a local approximation step. Some extensions have been added to efficiently cope with large-magnitude terms of opposite sign. Comparisons to other approaches have clearly indicated the efficiency and the appropriateness of the approximation algorithm.

Finally, the overall performance of the symbolic simulator ISAAC has been evaluated. It has been shown that ISAAC has the largest functionality for analog designers: it is dedicated to analog integrated circuits with a built-in small-signal linearization and a built-in expression approximation; it is also the only program which combines the symbolic analysis of both time-continuous and time-discrete circuits; it is also the only program which provides a symbolic distortion analysis. This distortion analysis will be discussed in the next chapter. It has also been shown experimentally that ISAAC's efficiency is better than or at least comparable to other approaches (as far as expanded expressions is concerned). Typical CPU times are 1.75 and 13 seconds for the voltage gain of a CMOS two-stage opamp and of a fifth-order switched-capacitor ladder filter, respectively.

It can be concluded that ISAAC presently outperforms all other tools for the symbolic analysis of analog integrated circuits. The program has proven itself to be a valuable design aid for experienced and novice designers, and is extensively being used in several universities and in industry. At the same time, the program is used to generate analytic models for automatically sizing the modules in our ASAIC analog design system presented in chapter 2.

All analyses in this chapter are restricted to linear characteristics of linear(ized) circuits. In the next chapter, the ISAAC program is extended towards the symbolic analysis of harmonic distortion in weakly nonlinear circuits. Also, several techniques will be presented to reduce the analysis time for larger circuits.

5

SYMBOLIC
DISTORTION ANALYSIS

5.1. Introduction

In the previous chapter, the algorithmic aspects of the symbolic simulation of analog circuits have been presented. The basic concepts have been illustrated for the symbolic simulator ISAAC [GIE_89c]. All analyses, however, were restricted to AC characteristics of linear(ized) circuits. In this chapter, several important extensions to the basic techniques are discussed such as for the symbolic analysis of harmonic distortion in weakly nonlinear analog circuits. In section 5.2, the symbolic analysis of noise behavior is discussed, in which an output variable is calculated for multiple inputs at once. In section 5.3, the symbolic analysis of harmonic distortion in weakly nonlinear circuits is discussed. An efficient analysis method is presented and implementation details are provided. In section 5.4, some other topics are discussed such as the symbolic calculation of sensitivity functions and symbolic pole/zero extraction. Techniques for the hierarchical analysis of large networks are then proposed in section 5.5. These techniques allow to reduce the CPU time for larger circuits. Finally, concluding remarks are presented in section 5.6.

5.2. Symbolic noise analysis

An important characteristic of analog integrated circuits is the noise behavior. Because of the small amplitudes involved, this noise behavior can be accurately calculated in the linearized circuit. Hence, the symbolic analysis techniques described in the previous chapter can be applied.

With respect to the **noise behavior of a circuit**, three different characteristics can be distinguished. The first one is the *transfer function* from a particular noise source to the circuit output (or calculated back to the input). The second and more important one is the *total output noise density* (or total equivalent input noise density), which consists of the contributions of all noise sources at the output (or input) as a function of the frequency. The third characteristic is the *total integrated output noise*, which is the total output noise density integrated over the frequency band of interest. This integration however is difficult to perform symbolically and provides little extra information. Besides, the total integrated output noise depends on the particular frequency band, and, for an operational amplifier for instance, on the feedback configuration applied to the amplifier.

The first two noise characteristics, on the other hand, can rather easily be calculated in symbolic form. Noise transfer functions are transfer functions as all others. ISAAC [GIE_89c] automatically adds the appropriate noise sources to the devices and calculates the noise transfer functions with the techniques presented in the previous chapter. The most important characteristic, however, is the total output noise density (which also depends on the feedback configuration e.g. for an operational amplifier) or the total equivalent input noise density (which is independent of the feedback configuration as far as the noise of the amplifier itself is concerned) [GRA_84]. **Symbolic expressions for the total noise density provide the analog designer with insight into the relative contributions of all noise sources over the whole frequency range.** In combination with the approximation algorithm presented in the previous chapter, only the dominant contributions at all frequency ranges are retained. Hence, it is rather straightforward to include symbolic noise analysis in the ISAAC program. Still, it requires some modifications. For the total (output or input) noise density, the output variable is calculated for multiple inputs (all noise sources) at once. Since in addition noise powers are summed, all noise transfer functions need to be squared before the summation, leading to very large expressions. An explosion of the number of terms, however, can be avoided by applying the approximation algorithm hierarchically, as will be described below. In the remainder of this section, first the device noise models used in ISAAC are reviewed shortly. Then, a method for the symbolic calculation and approximation of the total (output or input) noise density is described. Finally, a detailed example is given to illustrate the usefulness of the approach.

The noisy elements in analog integrated circuits are resistors, diodes and transistors. The **noise models** of a resistor, a bipolar and a MOS transistor are shown in Fig. 5.1 [GRA_84]. The noise of a resistor can be represented by a voltage noise source in series with the resistor or a current noise source in parallel with the resistor. Note that all noise sources in ISAAC are represented by a symbol. For example, the voltage noise source of resistor RX is denoted as DVN.RX, without substituting its value as $(4kT*RX)^{1/2}$. In a bipolar transistor, there is a noise source for each nonzero terminal resistance (such as the base resistance) and a current noise source between the collector and the emitter and one between the base and the emitter. These noise sources, however, are often combined into an equivalent input voltage noise source DVNEQ.QX and an equivalent input current noise source DINEQ.QX for bipolar transistor QX. Both sources consist of flicker noise and white noise. Note that these equivalent input noise sources DVNEQ.QX and DINEQ.QX are sometimes correlated and then better can be replaced by the real physical noise sources. In a MOS transistor, there also is a noise source for each nonzero terminal resistance and a current noise source between the drain and the source. Again, these noise sources are often combined into an equivalent input voltage noise source DVNEQ.MX for MOS transistor MX, consisting of flicker noise and white noise. Note also that the use of simple equivalent input noise sources is not exact anymore at really high frequencies because of the C_μ or the C_{GD} of the transistor.

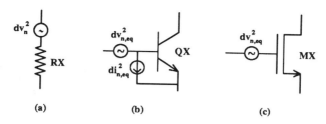

Fig. 5.1. Noise models in ISAAC for (a) a resistor, (b) a bipolar transistor, and (c) a MOS transistor.

In a practical circuit, several of these noisy elements, each with their own noise sources, are present. All noise sources have a propagation path to the output and the total noise density at the output is the sum of all contributions of all noise sources. Hence, the **total output noise density** dv_n^2 is given by :

$$dv_n^2 = \sum_i dv_{ni}^2 \star (TF_{ix})^2 \qquad (5.1)$$

where dv_n^i is the ith voltage (or current) noise source and TF_{ix} is the transfer function from this noise source to the circuit output x. Of course, the value of the output noise density is a function of the frequency. For the total equivalent input noise density, the above output noise density has to be divided by the square of the overall circuit transfer function from input to output (thereby canceling out the common system denominator).

Formula (5.1) can be calculated quite easily in ISAAC, since only the linear transfer functions TF_{ix} have to be derived. However, for practical circuits, these transfer functions are already lengthy and they still have to be squared in (5.1). This would lead to huge expressions in fully expanded form. Fortunately, all noise sources are uncorrelated. The contributions of any two sources do not interfere with each other. As a result, the approximation can safely be applied to each transfer function TF_{ix} separately before they are squared and summed, without causing any problems because of cancellations afterwards.

The **methodology for symbolic noise analysis** in ISAAC can then be described as follows. First, the transfer functions TF_{ix} are calculated and approximated. The noise expression (5.1) is then returned in nested format. In this format, all contributions of all noise sources are still present. Next, the noise expression is expanded by carrying out the polynomial squaring and summing, and again approximated. In this way, only the dominant contributions at all frequency ranges (or at a given center frequency) are retained in the final result, providing the analog designer with insight into the noise behavior of the circuit. Because of the hierarchical approximation, the CPU times are still acceptable. Note also that the technique of the adjoint network, which is often used for numerical noise calculations [DIR_69], requires exactly the same number of symbolic transfer functions to be calculated in ISAAC and hence shows no speed advantage above the straightforward calculation of the noise transfer functions (which have all lower-order minors in common). This is because Cramer's rule is used to calculate one output variable at a time in ISAAC [GIE_89c]. The adjoint network technique will however be advantageous if for instance symbolic Gaussian elimination is used. The symbolic noise analysis method is now illustrated in the following example.

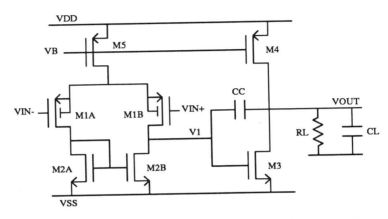

Fig. 5.2. CMOS two-stage Miller-compensated operational amplifier.

Example

Consider the CMOS two-stage Miller-compensated operational amplifier of Fig. 5.2. An equivalent input noise voltage is added to each MOS transistor. The exact expressions for the equivalent input noise density and for the total output noise density in unity feedback are plotted in Fig. 5.3. The element values are taken from the standard CMOS two-stage opamp presented in appendix A.

The approximated symbolic expression for the output noise at lower frequencies is given by :

$$(DVNEQ.M1A^2 + DVNEQ.M1B^2) + (DVNEQ.M2A^2 + DVNEQ.M2B^2) \, \frac{GM.M2^2}{GM.M1^2} \qquad (5.2)$$

The output noise at lower frequencies is dominated by the noise contributions of the transistors M1A, M1B, M2A and M2B of the input stage. At these frequencies, however, this is mainly 1/f-noise as clearly shown in Fig. 5.3. The noise of the biasing transistor M5 is suppressed by the square of the common-mode rejection ratio. The noise of the transistors M3 and M4 in the output stage is suppressed by the square of the gain in the first stage and is thus negligible at lower frequencies. As the frequency increases, the loop gain decreases but the transfer function for the propagation of the noise of the input-stage transistors to the output remains flat up till the gain-bandwidth of 1 MHz. From then on, this noise contribution starts to decrease. On the other hand, the transfer function for the noise of the output-stage transistors has a zero at a frequency well below the gain-bandwidth :

$$z_1 = \frac{GO.M1B + GO.M2B}{2\pi \ (CC + CGS.M3)} \tag{5.3}$$

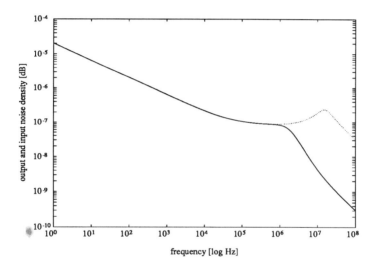

Fig. 5.3. Output noise density in unity feedback (straight line) and equivalent input noise density (dotted line) for the CMOS two-stage Miller-compensated opamp.

From this frequency on, the noise contribution of the second-stage transistors starts to increase and it reaches a maximum value at frequencies around the gain-bandwidth GBW. This maximum level can clearly be observed in Fig. 5.3 around 1 MHz. At higher frequencies, the noise again decreases due to the second pole p_2 in the opamp. The approximated symbolic expression for the output noise at frequencies around the gain-bandwidth is given by :

$$(DVNEQ.M1A^2 + DVNEQ.M1B^2) \ (\frac{GM.M1}{CC \ s})^2$$

$$+ \ (\text{DVNEQ.M2A}^2 \ + \ \text{DVNEQ.M2B}^2) \ \ (\frac{\text{GM.M2}}{\text{GM.M1}})^2 \ (\frac{\text{GM.M1}}{\text{CC S}})^2$$

$$+ \ \text{DVNEQ.M3}^2 \ + \ \text{DVNEQ.M4}^2 \ \ (\frac{\text{GM.M4}}{\text{GM.M3}})^2 \ (\frac{\text{CC} + \text{CGS.M3}}{\text{CC}})^2 \qquad (5.4)$$

◆

Conclusion

In this section, symbolic noise simulation has been discussed. The noise models of semiconductor devices as included in the ISAAC program have been discussed. It has been shown how the symbolic noise expressions can be derived in an efficient way, by carrying out the approximation hierarchically. Finally, the usefulness of symbolic noise analysis has been illustrated with the example of a CMOS two-stage opamp.

5.3. Symbolic analysis of harmonic distortion in weakly nonlinear analog circuits

In this section, techniques are discussed for the symbolic analysis of harmonic distortion in weakly nonlinear analog circuits. These techniques are implemented in the ISAAC simulator [WAM_90]. The importance of this analysis mode is shown in the following subsection.

5.3.1. Importance of symbolic distortion analysis

In the previous two chapters, only AC characteristics of linear(ized) circuits have been considered. This is, however, not sufficient for circuits such as filters, operational amplifiers and output buffers. Indeed, a very important characteristic of these circuits is the harmonic distortion, which is caused by the circuit and which degenerates the signal. Moreover, distortion is still poorly understood by the analog design community. Most of the designers only run some verifying simulations after completion of the design, and hope for a good measurement result. Very little is known of the different distortion sources and their relative contributions to the total harmonic distortion in real-life circuits [LAK_91]. Only few measurement data are available with respect to semiconductor device nonlinearities. The only well-known low-distortion design rule is the application of negative feedback to suppress distortion. The absolute value of the distortion in a circuit, however, is not easily predictable, and feedback cannot always be applied. Especially at higher frequencies, where capacitive effects become important and where the negative feedback usually is very small, the nonlinear behavior is very difficult to analyze. In addition, a whole category of circuits exists with an intrinsically nonlinear function, such as multipliers and mixers. They cannot be analyzed with a linear symbolic simulator, but require an extension towards the symbolic simulation of nonlinear circuits.

Numerical simulations with SPICE [VLA_80b] (e.g. transient analysis followed by Fourier transform for CMOS circuits) require very large CPU times and an optimal tuning of the simulation parameters (time step, accuracy, number of points...). With an inappropriate set of parameters, nearly any distortion figure can be obtained with SPICE in that way. Other numerical methods, for example based on Volterra series [BUS_74, CHU_79, SCH_80, VAN_83, and .DISTO in SPICE] or on the harmonic balance [KUN_86b] (currently used in the SPECTRE program [KUN_88]), can be more accurate in less CPU time. Still, they only provide a series of numbers. Of course, the contributions of the different nonlinearities can be shown individually in SPICE [VLA_80b], SWAP [SWA_83] or SPECTRE [KUN_88]. However, it takes many and long simulations to obtain a clear insight into the distortion generation and propagation mechanisms and to derive guidelines for low-distortion design. Therefore, the symbolic simulator ISAAC has been extended towards the symbolic analysis of harmonic distortion in weakly nonlinear, continuous-time analog integrated circuits [WAM_90]. It is interesting to mention that, up till now, ISAAC is the first and only program which generates symbolic expressions for the second and third harmonic and for the harmonic distortion of a circuit as a function of the fundamental frequency and the symbolic circuit parameters. All existing symbolic analysis programs only calculate linear characteristics. For practical circuits, the distortion expressions are again approximated based on the magnitudes of the circuit parameters. In this way, the major contributions to the harmonic distortion over the whole frequency range (or at a particular centre frequency) can be identified, providing the user with analytical insight into the nonlinear behavior of the circuit, which is complementary to the information obtained from numerical simulators.

Nonlinear circuits can be divided into circuits with hard nonlinearities and circuits with weak nonlinearities. The first category (for example a comparator) is hard to analyze symbolically. The second category can symbolically be **analyzed in the frequency domain with the Volterra series technique** [SCH_80], at least for relatively small signal levels. Fortunately, many circuits belong to this category :
- circuits which are intended to be linear, but which are to second order nonlinear because of nonidealities (e.g. operational amplifiers),
- circuits which are intended to be nonlinear but without hard nonlinearity (e.g. multipliers).

The Volterra series method then first calculates the first-order (linear) response of the circuit. The second-order response is then calculated as a correction on the first-order response. Next, the third-order response is calculated as a correction on the first-order and second-order responses, and so forth. This technique has been implemented in the ISAAC program [WAM_90].

For the symbolic distortion analysis, every nonlinear element is described with a power series expansion that is truncated after the first few terms. The basic nonlinear elements required for the distortion analysis of analog integrated circuits are described in subsection 5.3.2. The responses at the fundamental frequency, at the second harmonic and at the third harmonic are then calculated using the probing method [BUS_74, CHU_79]. Since this method generates complicated expressions, the length of which rapidly grows out of hand, a simplifying (but still exact) approach is necessary for symbolic analysis. It is described in subsection 5.3.3. The flowchart of the symbolic distortion analysis in ISAAC is then presented. To further reduce the number of terms and the CPU time, the expressions are approximated. However, much CPU time is gained by exploiting the mathematical structure of the distortion expressions and carrying out the approximation during the calculation of the harmonics. This requires an extension to the approximation algorithm described in the previous chapter and will be discussed in subsection 5.3.4. Finally, in subsection 5.3.5, some examples are presented, illustrating the feasibility and the usefulness of the approach.

5.3.2. Basic nonlinear elements and nonlinear transistor models

All devices commonly used in analog integrated circuits can for relatively small signals be represented by a set of four basic nonlinear elements, shown in Fig. 5.4: a conductance, a capacitor, a transconductance (controlled by one voltage) and a conductance controlled by more than one voltage. The first three elements can be linear and are part of the ISAAC primitives. The fourth element is inherently nonlinear and can be used to model for example the dependence of the collector current of a bipolar transistor on both the base-emitter voltage and the collector-emitter voltage. However, with only a small deviation at lower frequencies, the current can be approximated by a one-dimensional nonlinearity, thereby neglecting the distortion caused by the Early resistance. Indeed, in many cases, the low-frequency value of the distortion is not important, since it is suppressed by a large loop gain and thus reduced below the noise level. As a result, for a large category of circuits, the first three nonlinear elements of Fig. 5.4 are sufficient. Recently, however, the fourth nonlinear element has been included in the program as well to allow the simulation of circuits such as mixers and multipliers [WAM_91]. At the same time, the techniques have been generalized to the implementation of the analysis of multiple-input multiple-output systems.

All the basic nonlinear elements are voltage-controlled: the AC current flowing through the element can be written as a nonlinear function of one or more AC voltages. This is an interesting feature, since it makes the implementation of these elements in the CMNA method quite straightforward. For this, the nonlinear function which describes each element is expanded into a power series, which is then truncated after the third term. From now on, only the first three basic nonlinear elements are considered.

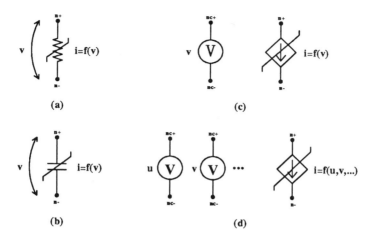

(a)   (c)

(b)   (d)

Fig. 5.4. Basic nonlinear elements for analog integrated circuits: (a) a nonlinear conductance, (b) a nonlinear capacitor, (c) a nonlinear transconductance, and (d) a nonlinear conductance controlled by more than one voltage.

The AC current through a nonlinear conductance or transconductance GX is then given by :

$$i = GX * v + K2.GX * v^2 + K3.GX * v^3 + \ldots \qquad (5.5)$$

where v is the controlling AC voltage, GX denotes the small-signal (trans)conductance of the linearized element, and K2.GX and K3.GX denote the second-order and third-order coefficients, respectively. Similarly, the AC current through a nonlinear capacitor CX is given by :

$$i = \frac{d}{dt} (CX * v + K2.CX * v^2 + K3.CX * v^3 + \ldots) \qquad (5.6)$$

where CX denotes the small-signal value of the linearized capacitor, and K2.CX and K3.CX denote the second-order and third-order coefficients. The other ISAAC rules for the notation of symbolic elements also remain valid.

For small signal levels, a bipolar and a MOS transistor can then be represented by an equivalent circuit consisting of several of the above basic elements. These models are shown in Fig. 5.5. They are generalizations of the linear small-signal models presented in chapter 3. For a bipolar transistor, five elements are nonlinear:

R_π, C_π, C_μ, GM and CCS. For a MOS transistor, four elements are nonlinear: GM, GMB, CSB and CDB. The output conductance and all terminal resistances are to first order considered to be linear. Formulas for the coefficients of these nonlinear elements are obtained by taking the derivatives of the transistor model I-V equations [WAM_90]. Note that the MOS transistor model is approximative, since the drain current in general is a function of three voltages (V_{GS}, V_{BS} and V_{DS}). In the model of Fig. 5.5b, all crossterms in the Taylor expansion of the function $I_D = f(V_{GS}, V_{BS}, V_{DS})$ are neglected, the output conductance is considered to be linear and the gate and bulk transconductances are considered to be unrelated nonlinearities.

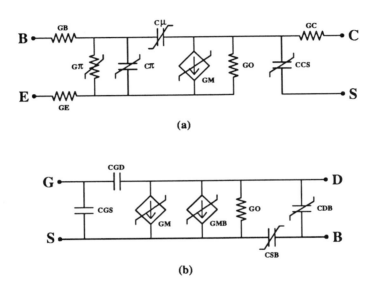

(a)

(b)

Fig. 5.5. Equivalent model for small signals for (a) a bipolar transistor and (b) a MOS transistor, with inclusion of nonlinear elements.

5.3.3. A method for the symbolic calculation of harmonic distortion

In this subsection, a method for the symbolic calculation of harmonic distortion is presented. For the sake of completeness, first the definitions of harmonic distortion are repeated here briefly. When a sinusoidal excitation A*cos(ωt) with amplitude A and frequency $f = \omega/2\pi$ is applied to a weakly nonlinear network, then the components of the voltage on any circuit node x at the fundamental frequency f, at the second harmonic 2f and at the third harmonic 3f are respectively given by :

$$V_x(\omega) = A |H_{1x}(j\omega)| \cos(\omega t + \alpha_{1x}) \tag{5.7}$$

$$V_x(2\omega) = \frac{A^2}{2} |H_{2x}(j\omega, j\omega)| \cos(2\omega t + \alpha_{2x}) \tag{5.8}$$

$$V_x(3\omega) = \frac{A^3}{4} |H_{3x}(j\omega, j\omega, j\omega)| \cos(3\omega t + \alpha_{3x}) \tag{5.9}$$

in which $H_{nx}(..)$ denotes the n-dimensional Fourier transform of the nth-order Volterra kernel or the nth-order transfer function of the node voltage, and α_{nx} is the argument of $H_{nx}(..)$. Note that the third-order effect at the fundamental frequency, as well as all higher-order contributions to the second and the third harmonic have been neglected here, which is only valid in the assumption of weakly nonlinear circuits. The second harmonic distortion ratio at node x is then given by the ratio of the component at $2f$ to the component at the fundamental frequency f :

$$HD_{2x} = \frac{V_x(2\omega)}{V_x(\omega)} \tag{5.10}$$

The third harmonic distortion ratio at node x is given by the ratio of the component at $3f$ to the component at the fundamental frequency f :

$$HD_{3x} = \frac{V_x(3\omega)}{V_x(\omega)} \tag{5.11}$$

Note that the harmonic distortion ratios HD_{2x} and HD_{3x} are complex variables with a magnitude and a phase, and that they depend on the frequency and on the amplitude of the applied input signal. The amplitude A is assumed to be small enough, such that higher harmonics can be neglected and that the second and third harmonic distortion ratios are sufficient to get an idea of the total harmonic distortion of the circuit.

The transfer functions $H_{1x}(j\omega)$, $H_{2x}(j\omega, j\omega)$ and $H_{3x}(j\omega, j\omega, j\omega)$ are determined using the **probing method** [BUS_74, CHU_79]. With this method, first the response of the linearized circuit to the input excitation at angular frequency ω is calculated. The second (third) harmonic is then calculated using the same linearized circuit, this time at the frequency 2ω (3ω). The inputs are now the nonlinear current sources of order two (order three) associated with each nonlinear element. These current sources form the second-order (third-order) corrections to the linear response, and are placed in parallel with the linearized equivalent of the corresponding nonlinear element. The value of these nonlinear current sources depends on the type of nonlinearity, on the order and on all lower-order responses [WAM_90].

The *nonlinear current source of order two* for a conductance GX between nodes p and q, or for a transconductance GX controlled by the voltage between nodes p and q is given by :

$$\text{INL2.GX} = \text{K2.GX} \ [H_{1p}(j\omega) - H_{1q}(j\omega)]^2 \qquad (5.12)$$

The nonlinear current source of order two for a capacitor CX between nodes p and q is given by :

$$\text{INL2.CX} = 2j\omega \ \text{K2.CX} \ [H_{1p}(j\omega) - H_{1q}(j\omega)]^2 \qquad (5.13)$$

Clearly, the nonlinear current sources of order two are a function of the first-order solution of the circuit nodes p and q. The nonlinear transfer function of order two at the circuit output x is then given by the response at this output due to all nonlinear current sources of order two :

$$H_{2x}(j\omega, j\omega) = \sum_i \text{TF}_{ix} \ \text{INL2}_i \qquad (5.14)$$

where TF_{ix} is the transfer function of the nonlinear current source of order two INL2_i to the circuit output x.

The *nonlinear current source of order three* for a conductance GX between nodes p and q, or for a transconductance GX controlled by the voltage between nodes p and q is given by :

$$\text{INL3.GX} = [H_{1p}(j\omega) - H_{1q}(j\omega)] \ *$$
$$\{2 \ \text{K2.GX} \ [H_{2p}(j\omega, j\omega) - H_{2q}(j\omega, j\omega)]$$
$$+ \ \text{K3.GX} \ [H_{1p}(j\omega) - H_{1q}(j\omega)]^2\} \qquad (5.15)$$

The nonlinear current source of order three for a capacitor CX between nodes p and q is given by :

$$\text{INL3.CX} = 3j\omega \ [H_{1p}(j\omega) - H_{1q}(j\omega)] \ *$$
$$\{2 \ \text{K2.CX} \ [H_{2p}(j\omega, j\omega) - H_{2q}(j\omega, j\omega)]$$
$$+ \ \text{K3.CX} \ [H_{1p}(j\omega) - H_{1q}(j\omega)]^2\} \qquad (5.16)$$

Clearly, the nonlinear current sources of order three are a function of both the first-order and the second-order solution of the circuit nodes p and q. The nonlinear transfer function of order three at the circuit output x is then given by the response at this output due to all nonlinear current sources of order three :

$$H_{3x}(j\omega, j\omega, j\omega) = \sum_i TF_{ix} \; INL3_i \qquad (5.17)$$

where TF_{ix} is the transfer function of the nonlinear current source of order three $INL3_i$ to the circuit output x. As a result, both for the calculation of the second and the third harmonic, the same linear circuit is analyzed but with different current source excitations. This explains why the techniques and algorithms in ISAAC, as described in the previous chapter for linear characteristics, can be applied to the symbolic analysis of harmonic distortion as well.

Inspection of the above formulas also shows that the symbolic expressions for nonlinear transfer functions can be quite huge. Fortunately, a simplifying (but still exact) approach is possible, since the denominator of all transfer functions can be factorized [WAM_90]. The denominator of a first-order transfer function H_{1x} is given by the determinant of the CMNA matrix of the linearized circuit $det(j\omega)$. The denominators of the second-order and third-order transfer function H_{2x} and H_{3x} are then given by $det(2j\omega)*(det(j\omega))^2$ and $det(3j\omega)*det(2j\omega)*(det(j\omega))^3$, respectively. Hence, once the determinant of the linearized circuit is known, only numerators of transfer functions need to be calculated, using only numerators of nonlinear current sources. For a nonlinear (trans)conductance GX controlled by the voltage between nodes p and q, the numerators of the nonlinear current sources of order two and three are given by :

$$INL2.GX = K2.GX \; [N_{1p}(j\omega) - N_{1q}(j\omega)]^2 \qquad (5.18)$$

$$INL3.GX = [N_{1p}(j\omega) - N_{1q}(j\omega)] \; *$$
$$\{2 \; K2.GX \; [N_{2p}(j\omega, j\omega) - N_{2q}(j\omega, j\omega)]$$
$$+ K3.GX \; det(2j\omega) \; [N_{1p}(j\omega) - N_{1q}(j\omega)]^2\} \qquad (5.19)$$

where $N_{px}(..)$ denotes the numerator of the pth-order transfer function of voltage x. For a nonlinear capacitor CX between nodes p and q, the numerators of the nonlinear current sources of order two and three are given by :

$$INL2.CX = 2j\omega \; K2.CX \; [N_{1p}(j\omega) - N_{1q}(j\omega)]^2 \qquad (5.20)$$

$$INL3.CX = 3j\omega \; [N_{1p}(j\omega) - N_{1q}(j\omega)] \; *$$
$$\{2 \; K2.CX \; [N_{2p}(j\omega, j\omega) - N_{2q}(j\omega, j\omega)]$$
$$+ K3.CX \; det(2j\omega) \; [N_{1p}(j\omega) - N_{1q}(j\omega)]^2\} \qquad (5.21)$$

These formulas clearly show that the symbolic distortion expressions still grow out of hand for practical circuits. This has two reasons. First, the number of nonlinear elements becomes quite large for circuits containing several transistors, since each transistor has four or five nonlinear elements. Each of these nonlinear elements gives rise to a nonlinear current source. Secondly, all these nonlinear current sources depend on the lower-order responses, whose symbolic expressions can already be quite large. For these reasons, an exact analysis of circuits of practical size would be impractical. However, the problem is solved by approximating the expressions before they are expanded. For this approximation, the same algorithm as described in the previous chapter is used. It again prunes the expressions based on the magnitudes of all circuit parameters. The main difference is that the mathematical structure of the distortion formulas is exploited and that the approximation now is carried out hierarchically during the calculation of the distortion expressions, as will be described in the next section. This strongly reduces the size of the expressions that are processed, and the CPU time. On the other hand, the hierarchical application of the approximation algorithm requires some special precautions in order to avoid large approximation errors. These will also be described in the next section.

The above discussion now results in the **ISAAC flowchart of the symbolic distortion analysis** depicted in Fig. 5.6. After the read-in and the expansion of the circuit and after the selection of a distortion analysis, the user has to supply the maximum approximation error. The CMNA matrix for the linearized circuit as well as the input vectors of order one to three are then constructed. At this moment, the nonlinear current sources of order two and three are still represented by a symbol $INL2_i$ or $INL3_i$, whose value will be calculated later on. Next, the linear response of the circuit to the input signal is calculated at the output and approximated. The transfer functions TF_{ix} for all nonlinear current sources to the circuit output are then calculated and approximated. This yields the nested expression (5.14) or (5.17) in which the nonlinear current sources are still represented by the symbols $INL2_i$ or $INL3_i$. This is the first possible output format for the distortion results in the ISAAC program. Next, the values of these nonlinear current sources of order two or three (depending on the requested analysis) are calculated. If at any moment during this calculation, a lower-order node voltage is needed which has not yet been calculated, it is calculated and stored for reuse later on. In this way, only the node responses are calculated which are really needed and they are calculated only once. For example, consider a nonlinear conductance between nodes p and q. Then during the calculation of the corresponding nonlinear current source of order two, the first-order (linear) solutions for the voltages on nodes p and q (if not the circuit output) are calculated. The values of the nonlinear current sources $INL2_i$ or $INL3_i$ are then approximated and substituted in the nested expression (5.14) or (5.17). This is the second possible output format for the distortion results in the ISAAC program. All contributions of all nonlinear elements are still present and clearly separated here. Finally, the obtained nested formulas (5.14) or (5.17) are expanded and approximated again. Nonlinearities with a small contribution to the output distortion are then discarded against nonlinearities with a larger contribution. In this way, only the really dominant contributions to the distortion at the output over the whole frequency range (or at the given center frequency) are retained. This is the third possible output format for the distortion results in the ISAAC program. All three formats will be illustrated in the examples of section 5.3.5.

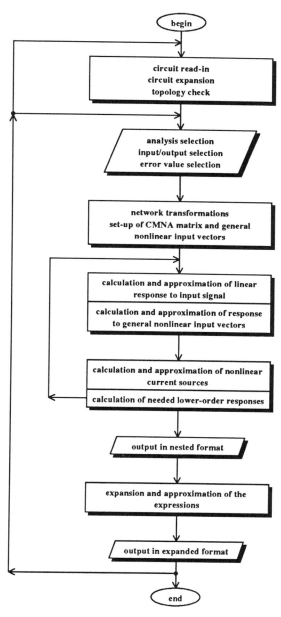

Fig. 5.6. Flowchart of the symbolic distortion analysis in ISAAC.

5.3.4. The approximation of symbolic distortion expressions

In the previous chapter, ISAAC's approximation algorithm has been discussed. It operated on the total, fully expanded expression. However, for symbolic distortion expressions, this approach is impossible due to the huge size of the fully expanded expressions. Fortunately, an efficient approximation is still practically feasible by exploiting the mathematical structure of the distortion expressions and performing the approximation hierarchically on different subexpressions before expanding the result. Care has to be taken, however, not to prune any terms which become dominant after expanding the results and after carrying out any remaining cancellations. Also, the terms present in the final result must be the most dominant ones and all terms must have the right numerical coefficient. These requirements are fulfilled by using the following technique.

Any second-order response can be written in the following form :

$$\sum_i K2_i * P_i * TF_{ix} \tag{5.22}$$

where $K2_i$ is the second-order coefficient of the ith nonlinearity, P_i is the polynomial which denotes the value of the corresponding nonlinear current source and which depends on the lower-order solutions of the two controlling nodes of this ith nonlinearity, and TF_{ix} is the transfer function of the nonlinear current source to the circuit output x. Hence, the corresponding polynomial products ($P_i * TF_{ix}$) can be approximated separately since terms in $K2_i$ cannot cancel with terms in $K2_j$ ($j \neq i$).

Similarly, any third-order response can be written in the following form :

$$\sum_i \{K2_i (\sum_j K2_j * P_j * TF_{ji}) + K3_i * P_i\} * TF_{ix} \tag{5.23}$$

It contains subexpressions in ($K2_i * K2_j$) and in $K3_i$. Each of these subexpressions are approximated separately, since they cannot cancel between each other anymore. The result is then expanded and approximated again to retain the dominant contributions only. In order to see the effective error introduced by the approximation, both the (nested) exact and the (expanded) approximated result can be evaluated numerically over some frequency range and then drawn in the same plot. The exact expression is of interest for numerical evaluations and for generating frequency plots. The approximated expression, however, is needed to provide the analog designer with analytical insight into the mechanisms of distortion generation and propagation in real-life circuits.

5.3.5. Examples of symbolic distortion analysis

The factorization and approximation presented in the previous two sections make symbolic distortion analysis possible with an acceptable amount of memory and CPU time, also for circuits of practical size. In this section, three examples are presented which show the feasibility and usefulness of the symbolic distortion analysis in ISAAC. The first example is a frequently used subblock: a CMOS current mirror. The second example is a bipolar emitter follower which is a circuit with internal distortion-suppressing feedback. The third example is a complete CMOS operational amplifier.

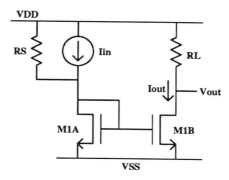

Fig. 5.7. CMOS current mirror.

Example 1
The first example is the CMOS current mirror, depicted in Fig. 5.7. The mirror is driven by a current source with impedance RS and loaded by a resistor RL. The two matching transistors M1A and M1B are considered to be nonlinear. The harmonic distortion is analyzed with mismatches of 1%. The 25%-error symbolic expression for the second harmonic distortion ratio at the output of the current mirror is given by :

```
AMP
 *  (       2 ΔGM.M1BA GL GO.M1 K2.GM.M1
          + ΔGM.M1BA GO.M1 GO.M1 K2.GM.M1
          + ΔGM.M1BA GL GL K2.GM.M1
      + S  (-2) GM.M1 GM.M1 GM.M1 K2.CDB.M1
      + S² (-8) CGS.M1 GM.M1 GM.M1 K2.CDB.M1
      + S³ (- 4 CDB.M1 CDB.M1 CGS.M1 K2.GM.M1
           + 16 CGD.M1 CGS.M1 GM.M1 K2.CDB.M1
           -  8 CDB.M1 CGD.M1 CGS.M1 K2.GM.M1
           -  2 CDB.M1 CDB.M1 CDB.M1 K2.GM.M1
           +  2 CDB.M1 CDB.M1 GM.M1 K2.CDB.M1)
      + S⁴ (- 8 CGD.M1 CGD.M1 CGS.M1 K2.CDB.M1
```

```
      - 4 CDB.M1 CDB.M1 CGD.M1 K2.CDB.M1
      - 12 CDB.M1 CGD.M1 CGD.M1 K2.CDB.M1))
```
$$\rule{6cm}{0.4pt}$$ (5.24)
```
(- GM.M1 + S CGD.M1) * 2 DET(2S) DET(S)
```

```
with DET(S) =       GO.M1 GM.M1 + GL GM.M1
               + S   CDB.M1 GM.M1
               + S² 2 CDB.M1 CGS.M1
```

where AMP is the amplitude of the input signal. This expression (5.24) clearly shows that in the ideal case, without mismatches, the current mirror does not introduce any harmonic distortion at lower frequencies. However, due to mismatches, the second-order nonlinearity of the transconductance of the MOS transistors does introduce some distortion at the output node. In addition, as the frequency increases, the nonlinearity of the drain-bulk junction capacitor becomes important and the second harmonic distortion ratio starts increasing (independent of any mismatches). This is also shown in Fig. 5.8, where the exact expression for the second harmonic distortion ratio is evaluated over some frequency range (scaled to an input current value of 1 A). The frequency in this figure is the fundamental frequency i.e. the frequency of the input signal. The low-frequency value is very small, since it depends on the second-order coefficient of a MOS transistor transconductance and on transistor mismatches, both of which are small. From about 0.1 Hz, the second harmonic distortion ratio then starts increasing. It reaches a maximum and then decreases again due to capacitive effects.   ◆

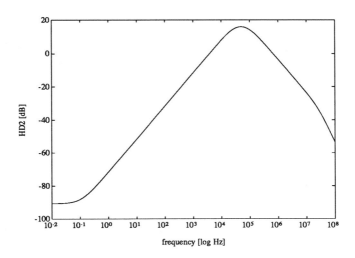

Fig. 5.8. Second harmonic distortion ratio at the output of the CMOS current mirror (scaled to an input current value of 1 A).

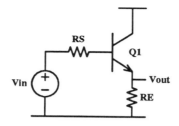

Fig. 5.9. Bipolar emitter follower.

Example 2

The second example is a bipolar emitter follower, shown in Fig. 5.9. It is interesting to investigate the distortion of this circuit, since it contains internal feedback, which as all negative feedback suppresses the distortion. The symbolic expression for the second harmonic distortion ratio at the output of the emitter follower with an approximation error of 25% is now given in three different formats. The first format is a nested expression, in which all nonlinear current sources of order two are still represented by a symbol :

```
AMP  *
(   INL2.Gπ.Q1
   *  (- GM.Q1 GO.Q1 + GO.Q1 GS + S 2 Cμ.Q1 GO.Q1)
  + INL2.Cπ.Q1
   *  (- GM.Q1 GO.Q1 + GO.Q1 GS + S 2 Cμ.Q1 GO.Q1)
  + INL2.GM.Q1
   *  (GO.Q1 GS + S 2 GO.Q1 Cπ.Q1)
  + INL2.Cμ.Q1
   *  (GO.Q1 GM.Q1 + S 2 GO.Q1 Cπ.Q1))
──────────────────────────────────────────────────────
GS * GO.Q1 * (GM.Q1 + S Cπ.Q1) * 2 DET(2S) DET(S)

  with DET(S) = GO.Q1 *
                (     GS (GM.Q1 + GE)
                 + S  (Cπ.Q1 GS + Cπ.Q1 GE + Cμ.Q1 GM.Q1)
                 + S² Cμ.Q1 Cπ.Q1)
```

$$(5.25)$$

This expression gives an idea of the influence of the different nonlinearities on the circuit output. In the second format, the values of the nonlinear current sources are substituted, but the formula is still kept nested :

```
AMP  *
(   K2.Gπ.Q1
   *  (GE GO.Q1 GS)
   *  (GE GO.Q1 GS)
   *  (- GM.Q1 GO.Q1 + GO.Q1 GS + S 2 Cμ.Q1 GO.Q1)
  + K2.Cπ.Q1
   *  (0 + S 2)
   *  (GE GO.Q1 GS)
```

```
     * (GE GO.Q1 GS)
     * (- GM.Q1 GO.Q1 + GO.Q1 GS + S 2 Cμ.Q1 GO.Q1)
  + K2.GM.Q1
     * (GE GO.Q1 GS)
     * (GE GO.Q1 GS)
     * (GO.Q1 GS + S 2 GO.Q1 Cπ.Q1)
  + K2.Cμ.Q1
     * (0 + S 2)
     * (- GS GO.Q1 GM.Q1 - GS GO.Q1 GE + S (-1) GS GO.Q1 Cπ.Q1)
     * (- GS GO.Q1 GM.Q1 - GS GO.Q1 GE + S (-1) GS GO.Q1 Cπ.Q1)
     * (GO.Q1 GM.Q1 + S 2 GO.Q1 Cπ.Q1))
```
$$\hrulefill \tag{5.26}$$
```
GS * GO.Q1 * (GM.Q1 + S Cπ.Q1) * 2 DET(2S) DET(S)

  with DET(S) = GO.Q1 *
                 (      GS (GM.Q1 + GE)
                  + S  (Cπ.Q1 GS + Cπ.Q1 GE + Cμ.Q1 GM.Q1)
                  + S² Cμ.Q1 Cπ.Q1)
```

This expression shows the contributions of all four nonlinearities to the harmonic distortion at the output and clearly illustrates the superposition of these contributions. However, to know the dominant distortion sources over the whole frequency range, the numerator of the above expression has to be expanded and again approximated, yielding :

```
AMP * GO.Q1 * GO.Q1 * GS *
(      GE GE GS K2.GM.Q1
  + S  (- 2 GE GE GM.Q1 K2.Cπ.Q1 + 2 GE GE GS K2.Cπ.Q1
        + 2 Cπ.Q1 GE GE K2.GM.Q1 + 2 GM.Q1 GM.Q1 GM.Q1 K2.Cμ.Q1
        + 4 GE GM.Q1 GM.Q1 K2.Cμ.Q1)
  + S² (4 Cμ.Q1 GE GE K2.Cπ.Q1 + 8 Cπ.Q1 GM.Q1 GM.Q1 K2.Cμ.Q1
        + 12 Cπ.Q1 GE GM.Q1 K2.Cμ.Q1)
  + S³ Cπ.Q1 Cπ.Q1 K2.Cμ.Q1 (10 GM.Q1 + 8 GE)
  + S⁴ 4 K2.Cμ.Q1 Cπ.Q1 Cπ.Q1 Cπ.Q1)
```
$$\hrulefill \tag{5.27}$$
```
(GM.Q1 + S Cπ.Q1) * 2 DET(2S) DET(S)

  with DET(S) = GO.Q1
                 * (      GS (GM.Q1 + GE)
                    + S  (Cπ.Q1 GS + Cπ.Q1 GE + Cμ.Q1 GM.Q1)
                    + S² Cμ.Q1 Cπ.Q1)
```

This expression now provides insight into the distortion behavior of the emitter follower. At lower frequencies the distortion is clearly dominated by the nonlinearity of the transconductance of the bipolar transistor. The second harmonic distortion ratio is then given by :

$$\frac{AMP}{2} * \frac{K2.GM.Q1}{GM.Q1} * \left[\frac{GE}{GM.Q1 + GE} \right]^2 \tag{5.28}$$

This expression can easily be interpreted: the distortion comes from the transconductance nonlinearity, which is given by K2.GM.Q2 times the square of the controlling base-emitter voltage. This controlling voltage is given by [GE/(GM.Q1 + GE)] v_{in}. From expression (5.28) low-distortion design rules can be derived. The ratio K2.GM.Q1/GM.Q1 is nearly constant for bipolar transistors. Hence, the larger the transconductance (and thus the current through the transistor) or the larger the resistor RE, the larger the loop gain (GM.Q1 * RE) and thus the smaller the second harmonic distortion ratio.

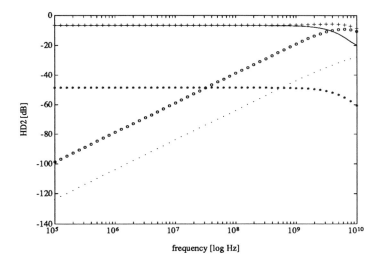

Fig. 5.10. Second harmonic distortion ratio at the output of the emitter follower, scaled to an input voltage of 1 V. The total harmonic distortion ratio (straight line) is shown, as well as the contributions of the individual nonlinearities: GM (+), Cπ (o), Gπ (*) and Cμ (.).

At higher frequencies the influence of the nonlinearities of Cπ and Cμ becomes more and more important. This is clearly shown in Fig. 5.10, where the exact expression of the second harmonic distortion ratio at the output of the emitter follower (scaled to an input voltage of 1 V) is evaluated over some frequency range. The distortion ratio remains flat up till 1 GHz, which is one decade below the f_T of the transistor for the set of values used in this example. From then on the distortion ratio decreases. Fig. 5.10 also shows the contributions of each individual nonlinearity. The distortion caused by Gπ is always smaller than the distortion caused by GM for this circuit. Similarly, the distortion caused by Cμ is smaller than the distortion caused by Cπ. In the region above 1 GHz, the nonlinearity of Cπ becomes as important as the nonlinearity of GM. However, since both contributions have an opposite sign at these frequencies, they cancel each other and the total harmonic distortion ratio becomes smaller than both individual contributions.

Similar expressions can now be derived for the third harmonic distortion ratio as well. The expressions however become more complicated since all second-order nonlinearities and all third-order nonlinearities contribute to the third harmonic distortion ratio. The value at lower frequencies for an approximation error of 25% is given by :

$$\frac{AMP^2}{4} \left[\frac{GE}{GM.Q1 + GE} \right]^3 \left[\frac{-2\ K2.GM.Q1^2 + (GM.Q1 + GE)\ K3.GM.Q1}{GM.Q1\ (GM.Q1 + GE)} \right] \quad (5.29)$$

The third harmonic distortion ratio is determined by both the second-order and the third-order nonlinearity of the transistor transconductance. Both contributions have an opposite sign and to some extent cancel each other. Fig. 5.11 shows the third harmonic distortion ratio at the output of the emitter follower (scaled for an input voltage of 1 V) over some frequency range. Note that this behavior is totally different from the one observed in Fig. 5.10 for the second harmonic distortion ratio.

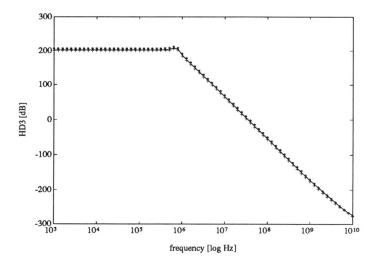

Fig. 5.11. Third harmonic distortion ratio at the output of the emitter follower, scaled to an input voltage of 1 V. The total harmonic distortion ratio is shown (straight line), as well as the contribution of the second-order (*) and third-order (+) coefficient of the nonlinear transistor transconductance.

Due to the efficient combination of nesting and approximation, the CPU time required for symbolic distortion analysis in ISAAC is quite low. The above expressions for the second and third harmonic distortion ratio at the output of the emitter follower have been derived in 0.09 and 0.57 seconds, respectively, on a TI EXPLORER II Plus.   ◆

In order to check the accuracy of the symbolic distortion analysis, the symbolic distortion expressions derived by ISAAC (both exact and approximated) have been compared to SPICE simulations (.DISTO command) [VLA_80b] for another bipolar circuit [WAM_90]. An extremely good agreement between both results has been observed (if the same set of values for the nonlinearity coefficients are used in both simulations). A similar comparison is not possible for CMOS circuits, since the .DISTO analysis is not implemented for MOS transistors in SPICE. ISAAC on the other hand can generate expressions for the harmonic distortion in CMOS circuits as well, as illustrated in the first and the following example.

Example 3

Consider now the CMOS two-stage Miller-compensated operational amplifier of Fig. 5.2. All transistor nonlinearities are modeled as shown in Fig. 5.5b. The symbolic expressions for the second and third harmonic distortion ratio at the opamp output in open loop are generated in 15.5 and 138.8 seconds of CPU time, respectively, on a TI EXPLORER II Plus. However, if all parasitic capacitances and thus all high-frequency effects are taken into account, the results are quite cumbersome and difficult to interpret, also for large approximation errors. More information can be obtained if the expressions are approximated around a given center frequency. For example, the 25%-error expression for the second harmonic distortion ratio at the output of the opamp in open loop is for lower frequencies given by :

$$- \text{AMP} \ * \ \frac{\text{K2.GM.M3}}{\text{GM.M3}} \ * \ \frac{\text{GM.M1}}{(\text{GO.M1} + \text{GO.M2})} \tag{5.30}$$

As one could expect, the largest distortion contribution is caused by the nonlinearity of the output transistor M3, which has the largest signal swing. Note that the other output transistor M4 cannot generate any distortion as it is driven by an ideal linear voltage source.

Another advantage of the symbolic simulator ISAAC is the possibility of carrying out a mixed symbolic-numerical analysis. The unapproximated distortion expression can be generated in nested format with all elements (except the complex frequency variable s) represented by their numerical value. The resulting expression in s can then efficiently be evaluated and plotted for several values of the frequency. A plot of the second harmonic distortion ratio in open loop at the output of the CMOS two-stage opamp (scaled to a differential input voltage of 1 V) is shown in Fig. 5.12. This plot clearly shows the variation of the distortion with the frequency of the input signal as well as the major contributions, and also allows to select a center frequency for generating any approximated, fully symbolic expression. The distortion is quite high at lower frequencies, but then decreases due to the influence of the compensation capacitor C_C. The internal feedback produced by this compensation capacitor reduces the distortion at the output. However, this effect vanishes at frequencies around the gain-bandwidth, which is 1 MHz for the above example. Fig. 5.12 clearly shows that the second harmonic distortion ratio in open loop is dominated by the second-order nonlinearity of the transconductance of the output transistor M3 over the whole frequency range up till the gain-bandwidth.

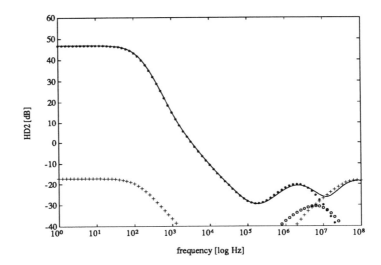

Fig. 5.12. Second harmonic distortion ratio at the output of the CMOS two-stage Miller-compensated opamp in open loop (scaled to a differential input voltage of 1 V). The total harmonic distortion ratio is shown (straight line), as well as the major contributions: $K2 . GM3$ (*), $K2 . GM1$ (+) and $K2 . GM2$ (o).

The situation is different if the same amplifier is connected in a feedback configuration. The open-loop distortion ratio $HD2_o$ is then divided by the square of the loop gain (if the same input voltage value is used in both configurations and if there is no feedforward through the feedback network). Hence, the closed-loop distortion ratio $HD2_c$ is given by :

$$HD2_c = \frac{HD2_o}{(1 + g\ A)^2} \qquad (5.31)$$

where g is the feedback factor and A is the open-loop gain, measured at the frequency of the second harmonic. The resulting second harmonic distortion ratio at the output of the CMOS two-stage opamp of Fig. 5.2 in unity feedback configuration (scaled to an input voltage of 1 V) and the major contributions are shown in Fig. 5.13. Due to the high loop gain at lower frequencies, the low-frequency value of the distortion ratio is now quite low. From about 1 kHz, the closed-loop distortion ratio starts increasing with increasing frequency, until a maximum value is reached at frequencies around the gain-bandwidth (1 MHz). Again, the harmonic distortion ratio is mainly determined by the second-order nonlinearity of the transconductance of the output transistor M3. However, capacitive feedforward of the nonlinear contributions of the input stage transistors (which is of course included in the ISAAC analyses) increases the harmonic distortion ratio at frequencies well below the gain-bandwidth. **This example clearly shows that a combination of symbolic expressions and exact numerical information is quite useful for a designer to obtain insight into the distortion behavior of complex circuits such as operational amplifiers.**   ◆

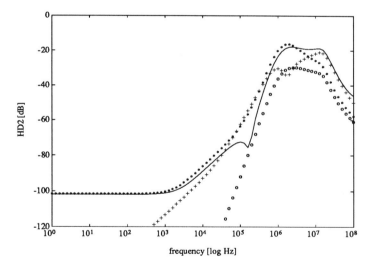

Fig. 5.13. Second harmonic distortion ratio at the output of the CMOS two-stage Miller-compensated opamp in unity feedback (scaled to an input voltage of 1 V). The total harmonic distortion ratio is shown (straight line), as well as the major contributions: K2.GM3 (*), K2.GM1 (+) and K2.GM2 (o).

Remark

It is important to realize that similar Volterra-series-based techniques can be used for the analysis of harmonic distortion in switched-capacitor circuits as well [VAN_83]. This has, however, not been implemented in symbolic form in the ISAAC program yet.

Conclusion

In this section, the extension of ISAAC towards the symbolic analysis of harmonic distortion in weakly nonlinear circuits has been discussed. The function which describes any nonlinear element is expanded into a power series and the first three coefficients are used for the symbolic analysis. For the distortion analysis of semiconductor circuits with a nominally linear behavior, only three nonlinear elements are sufficient, a resistor, a capacitor and a transconductance. These elements are included in the generalized small-signal models of bipolar and MOS transistors.

The probing method then allows to calculate the responses at the fundamental frequency, at the second harmonic and at the third harmonic by solving the same linear circuit but for different inputs. For the second and third harmonic analysis; the inputs are the nonlinear current sources associated to each nonlinearity. The flowchart of the symbolic distortion analysis in ISAAC has then been presented. In order to reduce the number of terms and the CPU time, the mathematical structure of the distortion expressions is exploited and the approximation is carried out during the calculation of the harmonics. Finally, several examples have been presented which illustrate the feasibility and the usefulness of the approach. These examples clearly show that ISAAC is a valuable tool for gaining insight into the complicated phenomenon of distortion in analog integrated circuits.

5.4. Symbolic sensitivity analysis and zero/pole extraction

In this section, two other topics are discussed, which are important for analog designers. In the first subsection, the generation of symbolic sensitivity expressions is considered. The generation of testability figures is mentioned briefly. In the second subsection, the symbolic extraction of zeros and poles from network functions is discussed.

5.4.1. Symbolic sensitivity analysis

An important characteristic for many analog circuits, such as filters, is the sensitivity of the circuit performance to variations in the values of the circuit parameters. These variations are unavoidable in practical circuit realizations due to process fluctuations, modeling and measurement inaccuracies, biasing and environmental (e.g. temperature) variations. As a result, because of these parameter variations, the performance of the circuit can deviate from the intended and simulated behavior. **The sensitivity is a measure for the variation in the circuit performance due to a variation in a circuit parameter**. In this context, two sensitivity figures have to be distinguished: the large-scale sensitivity and the small-scale sensitivity.

The **large-scale sensitivity** measures the variation in the circuit performance ΔP due to a large variation in a circuit parameter ΔX and is given by $(\Delta P / \Delta X)$. This information can be retrieved directly from the symbolic expression by generating the expression once and evaluating it for the different parameter value sets. Note that changes in the operating point are also taken into account if the magnitude values of the symbolic elements are adapted appropriately.

In practice, multiple parameters can vary at once according to some statistical model. The designer is then interested in the resulting statistical variation of the circuit performance. Usually, this is carried out by means of CPU-time expensive Monte-Carlo analyzes with a numerical simulator. An alternative and more efficient solution for some applications is to generate the symbolic expression once and then evaluate it repeatedly for different magnitude sets.

The **small-scale sensitivity** on the other hand measures the variation in the circuit performance δP due to an (in theory infinitesimally) small variation in a circuit parameter δX. Usually, relative variations are used and the small-scale sensitivity is then defined as :

$$S_X^P = \lim_{\Delta X \to 0} \left(\frac{\Delta P}{\Delta X} \frac{X}{P} \right) = \frac{\delta P}{\delta X} \frac{X}{P} = \frac{\delta \ln(P)}{\delta \ln(X)} \tag{5.32}$$

In the context of symbolic simulation, the circuit performance P in the above definition has to be replaced by some network function $H(s)$. The sensitivity with respect to any circuit parameter X can then be derived in symbolic form by symbolically differentiating the expression of the network function $H(s)$ to the circuit parameter X.

In practice, one is often interested in the sensitivity of the magnitude and the phase of $H(s)$ to variations in X. A straightforward relation, however, exists between the sensitivity of $H(s)$ and the sensitivities of the magnitude and the phase, as derived below :

$$S_X^H = \frac{\delta \ln(H)}{\delta \ln(X)} = \frac{\delta \ln(|H|e^{j\vartheta})}{\delta \ln(X)} = \frac{\delta \ln(|H|)}{\delta \ln(X)} + j \frac{\delta \vartheta}{\delta \ln(X)}$$

$$= S_X^{|H|} + j \vartheta S_X^{\vartheta} \tag{5.33}$$

where $|H|$ is the magnitude and ϑ is the phase of $H(s)$. If the parameter X is real, then the following relation holds for the sensitivity of the magnitude :

$$S_X^{|H|} = Re\{S_X^{H}\} \tag{5.34}$$

The sensitivity of $H(s)$ to a parameter X can easily be derived by symbolically differentiating the expression of $H(s)$. Also, for a rational function, the overall sensitivity is equal to the difference between the sensitivity of the numerator function and the sensitivity of the denominator function to the parameter X. Hence, the differentiation reduces to some kind of sorting between terms containing the parameter X and terms not containing the parameter X in both the numerator and the denominator [CHU_75]. This is now illustrated in the following example.

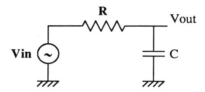

Fig. 5.14. Simple RC circuit to illustrate the symbolic sensitivity analysis.

Example
Consider the simple RC circuit of Fig. 5.14. The transfer function of this circuit is given by :

$$H(s) = \frac{V_{out}}{V_{in}} = \frac{N(s)}{D(s)} = \frac{1}{1 + R C s} \tag{5.35}$$

where $N(s)$ and $D(s)$ are the numerator and the denominator of the transfer function, respectively. The sensitivities of the transfer function $H(s)$ to the circuit elements R and C are then given by :

$$S_R^H = -S_R^D = \frac{-RCs}{1 + RCs} \tag{5.36}$$

$$S_X^H = -S_X^D = \frac{-RCs}{1 + RCs} \tag{5.37}$$

The sensitivities of the transfer function magnitude are then given by :

$$S_R^{|H|} = Re\{S_R^H\big|_{s=j\omega}\} = \frac{-R^2 C^2 \omega^2}{1 + R^2 C^2 \omega^2} \tag{5.38}$$

$$S_X^{|H|} = Re\{S_X^H\big|_{s=j\omega}\} = \frac{-R^2 C^2 \omega^2}{1 + R^2 C^2 \omega^2} \tag{5.39}$$

◆

These symbolic sensitivity expressions, however, can easily become cumbersome for practical circuits with large transfer functions, analog filters for example [LIB_89]. In addition, they have to be derived for every circuit parameter which can vary. Therefore, these sensitivity expressions can better be evaluated and plotted over some frequency range. This gives the designer a graphical idea about the influence of variations of the different parameters on the circuit performance as a function of the frequency. In this way, he can evaluate the quality of his design in terms of tolerance insensitivity or derive constraints for his design. If the design is not satisfactory, he can modify it and immediately see the results of any changes in (nominal) parameter values on the sensitivity by reevaluating the same expressions.

An alternative method to derive sensitivity functions is with the adjoint network approach [DIR_69]. This approach also allows to calculate small-scale sensitivities in symbolic form, by calculating circuit variables in both the original network and its adjoint network [CHU_75]. For example, for the above RC circuit, the current through the resistor R and the voltage across the capacitor C have to be calculated in the original network driven by a voltage source at the input and in the adjoint network (which is the same as the original network for this example) driven by a current source at the output (with the input shorted). The same sensitivity formulas are then obtained. However, the adjoint network approach has no advantage over the straightforward symbolic differentiation method in ISAAC since for each variable two network functions have to be calculated in symbolic form. The technique of the adjoint network can be useful for deriving numerical sensitivity figures, since the sensitivities to all parameters are derived at once by numerically analyzing the original and the adjoint network. However, these two analyses have to be performed at each frequency point. Therefore, the method of first generating the symbolic sensitivity functions and then evaluating these expressions for each set of parameter values and for each frequency point can be more efficient if the number of frequency points is large and the number of terms in the expressions not excessive [CHU_75].

Finally, in [LIB_89], a method is mentioned for the testability analysis of linear networks. This method, which is included in the SAPEC program, returns a testability figure, which indicates the number of testable (or observable) elements for a given set of test points. The method depends on the number of linearly independent rows in the Jacobian of the circuit and is speeded up by the use of symbolic sensitivity functions.

5.4.2. Symbolic calculation of zeros and poles

For the design of analog circuits, it is important to know the location of the poles and zeros of the circuit. For example, for filters these poles and zeros determine the filter characteristic. For operational amplifiers, the poles and zeros determine the stability of the circuit. Therefore, it is important for a designer to obtain symbolic expressions for the poles and zeros of a network function.

The symbolic simulator ISAAC [GIE_89c] generates fully symbolic expressions for a network function over the whole frequency range. However, up till now, no general and robust method exists yet to symbolically extract the zeros and poles out of this network function. Only some approximate methods exist which hold for some specific cases. The most widely used approximation is the **pole-splitting hypothesis** [GRA_84]. Consider for example the following polynomial function :

$$A_0 + A_1 \ s + A_2 \ s^2 + \ \ldots \tag{5.40}$$

If one can assume that the first zero is largely separated from the other zeros in this polynomial, i.e. if $A_{12} \gg A_0 \ A_2$, then the first zero is given by :

$$z = - \ \frac{A_0}{A_1} \tag{5.41}$$

This zero can now easily be split off and one can apply the same technique on the remaining polynomial function.

However, if the pole-splitting hypothesis fails, no general method exists yet to extract the remaining zeros. And it is not surprising that this situation often happens in practical circuits. For example, the first pole of an operational amplifier in open loop is situated at low frequencies and can easily be split off. All other poles and zeros, however, are lying close to each other at frequencies above the gain-bandwidth and the pole-splitting hypothesis fails. Therefore, the symbolic extraction of zeros and poles has not been included in the ISAAC program. On the other hand, a designer can usually write down approximate expressions for the poles and zeros of a circuit by exploiting topological information, i.e. by assigning the different poles to different circuit nodes [SAN_88b]. But this assumption is not general either, especially not when the poles become complex. Therefore, the symbolic extraction of zeros and poles of network function is still a field for future research. In addition, one must always keep in mind that the location of poles and zeros can seriously be affected by any approximation carried out on the symbolic expressions.

Conclusion

In this section, the generation of symbolic sensitivity functions has been discussed. It has been shown that these functions can easily be generated in symbolic form, but that they are mainly useful as intermediate results for further numerical and graphical processing. With respect to the symbolic extraction of zeros and poles of a network function, no general method exists yet and only some approximate techniques, such as the pole-splitting hypothesis, can be applied.

5.5. Techniques for the hierarchical symbolic analysis of large circuits

In the previous chapter, the limitations of symbolic simulators such as ISAAC with respect to the maximum analyzable circuit size have been pointed out. These limitations mainly result from the large number of terms in the expressions, which cause the CPU time and the storage requirements to increase exponentially with the size of the circuit, at least if the expressions are generated in expanded format. This expansion, however, is required by the expression approximation algorithm. As a result, for really large circuits, the symbolic simulation fails when the physical memory of the computer is full.

To extend the capabilities of symbolic simulation to larger circuits, a few techniques are possible. They are described here shortly, as a suggestion for future research.

5.5.1. Nested formulas and hierarchical decomposition

Obviously, the only way to overcome the circuit size limitations is to avoid the expansion of the expressions, to keep the expressions nested and to carry out the approximation during the network function calculation if possible. The nesting of expressions has several advantages, such as the reduction of the CPU time and the storage requirements, but also the reduction of the number of operations needed to evaluate the expression numerically.

The nesting of the expressions can be introduced in two different ways.

- First of all, **the determinant can be calculated in nested form**. The sparse recursive Laplace expansion algorithm can still be applied, but the polynomial multiplications and summations are not carried out anymore and are included as an operator in the expression. This will considerably speed up the calculation time. For example, the calculation of the differential gain of a folded-cascode OTA requires 0.32 s in nested format and 161.6 s in expanded format. The FDSLEM algorithm presented in [SMI_81b] could be useful in this context. On the other hand, the individual coefficients of the frequency are not available anymore and a new reliable approximation algorithm has to be developed.

- Secondly, the **nesting can be introduced by hierarchically decomposing the circuit** [STA_86]. The nested calculation of the determinant, as explained above, has no direct relation to the circuit topology. The topological information is exploited much better if the circuit is partitioned in subcircuits, these subcircuits eventually in subsubcircuits, etc. as shown in Fig. 5.15a. This hierarchical decomposition can then be represented by a decomposition tree as shown in Fig. 5.15b.

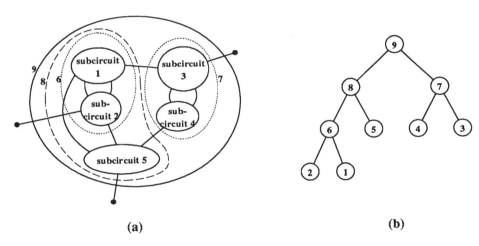

(a)   (b)

Fig. 5.15. Hierarchical decomposition (partitioning) of a circuit and corresponding decomposition tree.

To obtain the symbolic network function for the total circuit, first all lowest-level subcircuits are analyzed separately. Next, these subcircuits are combined on a higher level. The expression for a higher-level (sub)circuit, formed by interconnecting several lower-level subcircuits, is generated by combining the expressions of the composing subcircuits in a nested way. Note that input and output impedances of the different subcircuits (loading effect) become important here. This combination process is repeated up the hierarchy until the highest level is reached and the expression for the total circuit has been generated. The CPU time required for the analysis now depends on the type of partitioning or decomposition used. For a cascade connection of subcircuits, the CPU time and memory requirements increase linearly with the size of the network [STA_86]. When the connection of the subcircuits is more complicated than cascade, the analysis can be more time consuming.

Therefore, for an efficient symbolic analysis of large-scale circuits, the number of interconnection nodes between each two subcircuits has to be minimized. For active RC filters for instance, the circuit can be partitioned at every opamp output. In general, however, **the problem is to find an optimal partitioning of the circuit nodes**, such that nodes within the same group are or can be strongly interconnected but that the number of interconnections between the different groups of nodes are minimized. The selection of such an optimal partitioning, however, is an NP-complete problem. An efficient heuristic partitioning algorithm which yields near-optimal solutions can be found in [SAN_77].

The nesting introduced by such hierarchical decomposition reflects the circuit topology in a more natural way than simply calculating the determinant in nested form, since strongly interconnected nodes are analyzed in the same subcircuit, whereas the different subcircuits are only weakly interconnected. For example, a two-stage opamp can be partitioned in a first stage and a second stage. Both stages are analyzed seperately and are then combined to obtain the transfer function of the complete opamp. This can also be considered as some kind of high-level simulation, since each subcircuit (opamp stage) is first represented as a two-port, characterized by its own transfer function, input and output impedance. Next, the interconnection between these two-ports is simulated at the opamp level in terms of the transfer function, input and output impedance of each two-port. In general, however, a subcircuit can have more than two ports and multi-port parameters have to be used.

The techniques of nested formulas and hierarchical decomposition are expected to make the symbolic analysis of large circuits possible. Hierarchical decomposition has already been used in combination with a signal-flow-graph method in [STA_86], and in combination with a matrix-based method in [HAS_89a]. However, both programs generate exact expressions only. This can be useful for filter structures, but for circuits such as operational amplifiers an approximation is needed. Unfortunately, up till now, no efficient and reliable approximation algorithm exists yet for expressions in nested format. Therefore, the expressions still have to be expanded in the present version of ISAAC [GIE_89c, SAN_89]. **Future research has to concentrate on methods for approximating during the expression calculation (if possible) or on methods for approximating the nested expression after it has been calculated (without complete expansion).** The development of hierarchical approximation or pruning algorithms is an important problem which has to be solved for the symbolic simulators of the next generation. In addition, together with the nesting of the expressions, problems can be expected with remaining term cancellations, especially for reduced circuit equation formulations such as the MNA and CMNA methods. It might be that, for the generation of the network functions in nested format, other formulations are more appropriate. This has to be investigated as well before developing any new symbolic simulator.

5.5.2. Rules and the exploitation of topological knowledge

Another technique to reduce the CPU time and the expression complexity for large circuits is to cut at the modeling side of the symbolic simulator. **The small-signal equivalent circuit, which is used for generating the symbolic expressions, can often be simplified without introducing any noticeable error**: the resulting expressions after approximation are the same over a large frequency range. The simplification of the small-signal circuit is then based on the exploitation of topological knowledge about the circuit and its behavior.

Several possible simplifications are now described shortly.
- **A first possibility is to skip some small elements out of the circuit model.** For example, for a MOS transistor in saturation, the gate-drain capacitance consists of the small oxide overlap capacitance only and can be neglected (if the Miller effect is small), without significant effect on the major poles and zeros of the circuit.
- **Secondly, the circuit can be simplified by exploiting topological information.** For an opamp for example, the biasing circuit can be replaced by (ideal) current and/or voltage sources without changing the signal behavior of the circuit.
- **A third topological simplification is the exploitation of symmetry in the circuit.** For a symmetric circuit, such as a fully differential-in differential-out opamp, the two symmetrical subcircuits can be analyzed separately with the edges connecting both subcircuits left open if the circuit is driven with a common-mode voltage, or with the edges connecting both subcircuits shorted to AC ground if the circuit is driven with a differential-mode voltage. This is the well-known bisection theorem, which holds for symmetric circuits only.
- **Finally, heuristic rules can be used as well to impose some simplifications under certain conditions.** For example, even if the circuit is not perfectly symmetric, the above bisection theorem can still be applied in an approximate way. A typical example is the common-source (or common-emitter) voltage of a differential pair driven by a differential input voltage. Without noticeably changing the circuit behavior, this voltage can be considered as AC ground.

The above techniques can be applied to simplify the circuit schematic and thus the analysis time and the complexity of the resulting expressions, without noticeably changing the behavior of the circuit. It is important to note that these techniques can already be applied in the present version of ISAAC, but the simplifications are not yet introduced in an automatic way.

Conclusion

In this section, several techniques have been proposed to extend the capabilities of symbolic simulation to larger circuits. The final circuit size aimed at with symbolic simulation is of the order of practical building blocks, such as entire amplifiers, output buffers, leapfrog filters, etc. To achieve this, a nested representation of the expressions and an hierarchical decomposition of the circuit are necessary. However, at present, the problem with these methods is the lack of an efficient and reliable hierarchical approximation algorithm. Also, several techniques have been presented to simplify the AC circuit schematic before carrying out the symbolic analysis, by exploiting topological information about the circuit.

5.6. Conclusions

In this chapter, the basic techniques for the linear symbolic analysis of analog circuits presented in the previous chapter have been extended. A first extension is the symbolic analysis of noise behavior, which can efficiently be performed by carrying out the approximation of the expressions hierarchically. A second extension is the symbolic analysis of harmonic distortion in weakly nonlinear analog circuits. An efficient method, derived from Volterra series theory, has been presented which allows to calculate the responses at the fundamental frequency, at the second harmonic and at the third harmonic by solving the same linear circuit, but for different inputs. For the second and third harmonic analyses, the inputs are the nonlinear current sources associated to each nonlinearity in the circuit. Several practical examples illustrate the feasibility and the usefulness of the approach, which provides designers insight into the complicated phenomenon of distortion in analog integrated circuits.

Next, the symbolic calculation of sensitivity functions has been discussed, but these expressions are mainly useful as intermediate results for further numerical and graphical postprocessing. With respect to the symbolic extraction of zeros and poles of a network function, no general method exists yet and only some approximate methods, such as the pole-splitting hypothesis, can be applied. Finally, some techniques have been suggested to extend the capabilities of symbolic simulation to larger circuits, such as hierarchical decomposition, the use of rules and the application of topological knowledge. For large circuits, it is necessary to calculate the expressions in nested format, but up till now no efficient and reliable hierarchical approximation method exists yet.

The many examples presented in this chapter clearly show that a symbolic simulator such as ISAAC is a valuable and reliable tool to aid analog designers in understanding the behavior of analog integrated circuits. In addition, since the symbolic expressions accurately predict the circuit behavior, they can as well be used during the design of a circuit. Therefore, in the next chapter, the automatic sizing of analog integrated circuits based on analytic circuit models is discussed.

This then illustrates how symbolic analysis techniques can be used to create an open analog design system.

6 ANALOG DESIGN OPTIMIZATION BASED ON ANALYTIC MODELS

6.1. Introduction

In this chapter, it is illustrated how the symbolic analysis techniques presented in the previous three chapters can be used to create an open analog design system and to size practical circuits. In gneral, the design of analog integrated circuits starting from specifications and technology data can be divided in three major tasks: the selection of an appropriate schematic, the sizing of this schematic and the generation of the layout. Sizing (or dimensioning) is the determination of all bias parameters, element values and device sizes in the circuit in order to satisfy all performance constraints, while possibly also optimizing some design objectives. It is a critical step in the overall design process, since many high-performance applications and mixed analog-digital ASICs require an optimal design solution for the analog circuits in terms of area, power and overall performance. Therefore, since the specifications and thus the optimal device sizes can largely vary from application to application, the analog circuits have to be sized dedicatedly for each application instead of using fixed cells.

In chapter 2, the ASAIC system for the automated design of analog functional modules has been presented. The sizing in ASAIC is generally formulated as a constrained optimization problem, based on an analytic description of the circuit behavior and simulated annealing. It is implemented in the analog optimization program OPTIMAN (OPTIMization of Analog circuits) [GIE_89a, GIE_90b]. To speed up the optimization, all circuits are characterized by an analytic model, which consists of a large set of equations, describing the circuit's performance and additional constraints, as well as a set of independent variables. All AC characteristics in this analytic model are automatically generated by means of the symbolic simulator ISAAC, which has been described in the previous three chapters, enabling a fast inclusion of new topologies. The equations of the analytic model are then manipulated by the DONALD program [SWI_90] in order to determine a set of independent variables and to transform the equations into an appropriate solution plan for the optimization. Based on this solution plan, OPTIMAN then sizes all elements to satisfy the performance constraints, thereby optimizing a user-defined design objective. The global optimization method being applied on the analytic circuit models is simulated annealing.

The next section 6.2 generally discusses the requirements for any sizing program. The two different approaches to solve the underconstrained sizing problem are compared and the motivation for an optimization-based method relying on analytic models is shown. Section 6.3 then describes how analog design within OPTIMAN is formulated as a constrained optimization problem, and briefly discusses the major algorithmic aspects of the program. Section 6.4 then presents several practical examples and experimental results, and compares the performance of the program to other approaches from the literature. These examples show that OPTIMAN quickly designs analog circuits which closely meet the specifications, and that it is a flexible and reliable design tool. Finally, to complete the analog design cycle, section 6.5 briefly describes the major issues in automated layout generation for analog integrated circuits. Concluding remarks are then provided in section 6.6.

6.2. Circuit sizing based on an optimization of analytic models

In recent years, several programs have been published for the dimensioning of analog integrated circuits [ALL_83a,84,85,86b, BER_88a,89, BOW_85, CAR_88c, DEG_84,87c,89,90,   ELT_86,87,89,   FUN_88,   GIE_89a,90b,   GOF_89a, HAR_87,88ab,89b,   HAS_89b,   HEI_88,89,   HSU_89,   JUS_90,   KEN_88, KOH_87,90, LAI_88, NYE_81,88, OCH_89, ONO_89,90, POR_89, RAA_88, SAS_89, SHE_88, TOU_89, WAL_91]. In each of these programs, the analog circuits are resized for every application instead of using fixed cells. This guarantees an optimal solution in terms of area, power and overall performance, especially for ASICs. After an appropriate circuit schematic has been selected from the program's library (either by the user or by the program, either hierarchically or flat), the program sizes all circuit elements, to satisfy the performance constraints, possibly also optimizing an objective function. A similar approach is adopted in the ASAIC design system, which has been presented in chapter 2. Also, to perform the sizing, all programs use some analytic description (design equations) of the circuit and exploit expert designer knowledge to some extent.

The optimization strategy adopted in the OPTIMAN program is motivated in this section. In the first subsection 6.2.1, the complexity of analog circuit sizing is shown, and optimization-based methods are compared to knowledge-based methods. Next in subsection 6.2.2, the basic idea of optimization based on analytic models is introduced. An appropriate optimization algorithm is then selected in subsection 6.2.3.

6.2.1. Optimization-based versus knowledge-based sizing

The sizing (or dimensioning) of a circuit is the determination of the circuit parameters (bias currents and voltages, element values, device sizes) from the imposed specification values for the given technology and process data. If we denote the set of circuit parameters as $\underline{x}$ and the set of circuit performance characteristics (to be distinguished from the specifications) as $\underline{p}$, then the circuit performance for a given set of circuit parameters or the mapping from $\underline{x}$ to $\underline{p}$ can easily be determined by means of analysis (either by analytic equations or by numerical simulation). As shown in Fig. 6.1, sizing is the determination of the circuit parameters for given performance specifications, or the inverse mapping from $\underline{p}$ to $\underline{x}$. This is based on an inversion of the analysis equations into design equations. Usually, this inverse mapping is not unique: the number of design equations in practical designs is much smaller than the number of variables. Hence, the problem is underconstrained with many degrees of freedom. Many different solutions possibly satisfy the same specifications, but the design can be unrealizable as well.

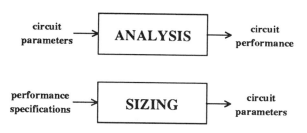

Fig. 6.1. Definition of sizing (or dimensioning).

In order to obtain a good design solution, the degrees of freedom have to be eliminated while optimizing some design objectives. In the present analog design systems, this is accomplished in essentially two different ways: either by exploiting analog design knowledge and heuristics, or by optimization.

1. Knowledge-based sizing

Knowledge-based methods explicitly use the existing analog design knowledge and heuristics to eliminate the degrees of freedom in the design and to obtain a feasible solution in a rather straightforward way. Typical examples are IDAC and OASYS. In IDAC [DEG_87c], the circuit equations are solved explicitly by experienced designers to yield a straightforward step-by-step design procedure, called a formal description, for each transistor schematic, which is then executed during the sizing. OASYS [HAR_89b], on the other hand, is a knowledge-based synthesis framework which uses design plans to successively select topologies and translate specifications downwards in the hierarchy until the transistor level is reached.

2. Optimization-based sizing

Optimization-based methods eliminate the degrees of freedom in the design by considering the sizing as a constrained optimization problem with some objective functions and some boundary conditions. Typical examples are DELIGHT.SPICE

[NYE_88] and OPASYN [KOH_90], but this approach is also adopted in the OPTIMAN program [GIE_90b].

Knowledge-based programs have the advantage of exploiting the existing design knowledge and hence obtaining a solution in a very fast way. However, the knowledge and heuristics are very difficult to formalize in a general and context-independent way. Different designers often have different strategies and different rules of thumb. Due to the many heuristics, the resulting design is just an initial first-cut solution which still has to be fine-tuned by some local optimization method. In addition, the heuristics used, for example the circuit-specific formal procedures in IDAC [DEG_87c] or the design plans in OASYS [HAR_89b], are always oriented towards some implicit optimization goal, such as minimum power and area, and are thus less suited for other optimization goals, such as minimum noise. It is also difficult to adequately manage multiple conflicting requirements and design trade-offs with simple heuristics. As a result, the heuristically obtained initial design can be far from the global optimum solution, such that even a fine-tuning by local optimization cannot obtain the best overall solution.

Optimization-based methods, on the other hand, inherently take care of all conflicting requirements and design trade-offs. Since the optimization goals are stated explicitly and are controlled by the user (or the design system), these methods allow a general optimization of the circuit targeted to each actual application. Also, with an appropriate optimization algorithm, a solution close to the global optimum can be obtained for any application. Therefore, in the ASAIC design system, an optimization-based method is adopted. It is implemented in the analog optimization program OPTIMAN (OPTIMization of Analog circuits) [GIE_89a, GIE_90b]. However, the drawback of all optimization-based methods is that they require more CPU time than knowledge-based methods due to the many iterations. **The difference between both methods is to some extent the trade-off between design flexibility and CPU time.**

In this subsection, the selection of an optimization-based method has been motivated. In the next subsection, the idea of optimization based on analytic models is introduced.

6.2.2. Optimization based on analytic models

A classification of sizing methods and more specifically optimization-based methods is shown in Fig. 6.2. A straightforward approach to analog circuit design optimization is the combination of algorithmic optimization and numerical simulation, as in DELIGHT.SPICE [NYE_88]. This approach is applicable to a very broad range of analog circuits and produces near-optimal results. However, it is inefficient and costly in terms of CPU time, due to the full simulation at every iteration. On the other hand, an optimization program dedicated to one particular circuit schematic, such as BISON [WAL_87], can be very fast, but is applicable to that schematic only.

- **knowledge-based sizing**
    Ex: IDAC, OASYS, BLADES
- **optimization-based sizing**
    dedicated to a topology
        Ex: BISON
    numerical simulation in loop
        Ex: DELIGHT.SPICE
    evaluation of analytic models in loop
        Ex: OPTIMAN, OPASYN

Fig. 6.2. Classification of sizing methods.

In order to combine the advantages of optimization-based methods with acceptable CPU times, the following alternative is adopted in the OPTIMAN program [GIE_89a, GIE_90b]. The full numerical simulation at every iteration is replaced by the evaluation of an analytic model, consisting of design equations modeling the circuit's behavior. This approach is based on the observation that most circuit characteristics, such as the gain or the phase margin of an operational amplifier, are influenced by a small number of parameters only. All other circuit parameters have only a negligible influence, even over a broad range of values. As shown in chapters 3 and 5, simplified expressions can indeed be a good approximation for the circuit behavior over the whole frequency range. Hence, one can accurately model the circuit behavior by means of such simplified analytic equations. This has two advantages. First, the evaluation of analytic equations highly speeds up the optimization by reducing the time spent at every iteration. In this way, an excellent trade-off between CPU-time consumption during the optimization and accuracy is obtained. Secondly, the derivation of the analytic models can largely be automated. In OPASYN [KOH_89, KOH_90], the same approach of optimization based on analytic models is used. However, whereas in OPASYN the analytic models for a new topology must first be created by a good analog designer, which is a tedious and error-prone job, the analytic models for OPTIMAN are generated in a highly automated way by means of the symbolic simulator ISAAC, which has been discussed in the previous chapters. This allows the fast and error-free inclusion of new topologies into the design system and does not restrict the system to a fixed set of topologies, which is one of the major limitations of most existing design systems, especially for fully knowledge-based systems.

For any actual application, the equation manipulation program DONALD [SWI_90] then automatically transforms the context-independent analytic models into a form which is suitable for the optimization. To this end, DONALD selects a set of independent variables and creates a dedicated solution plan for the optimization. Based on this solution plan, OPTIMAN then sizes the circuit topology to satisfy all performance specifications. The remaining degrees of freedom are used to optimize a general user-tunable cost function. After OPTIMAN returns the optimized circuit parameters, the design is verified in ASAIC. The use of simplified analytic models speeds up the optimization, but the resulting design has to be checked afterwards with more accurate numerical simulations. If some specifications are not satisfied, the redesign system is entered (see chapter 2) and possibly the circuit has to be redesigned. If no acceptable solution can be found during the sizing, OPTIMAN also indicates which characteristics are responsible for the failure.

6.2.3. Selection of an optimization algorithm

The last problem for an optimization-based method is the selection of an appropriate optimization algorithm [BRA_81]. This algorithm must be quite general and applicable to a wide range of diverse applications, such as the sizing of opamps or the optimization of switched-capacitor filters. It is likely possible that some specialized algorithms are mathematically more appropriate for certain specific applications. However, instead of selecting a different optimization algorithm for each application from a box of possible algorithms, **for the OPTIMAN program it has been decided to implement one general optimization algorithm which can obtain a global solution for many applications in reasonable CPU times.** At the same time, the number of control parameters of the optimization algorithm, which are to be controlled by the user and which determine the quality of the resulting solution, must be as small as possible. The problem of sizing a circuit must not be replaced by the problem of selecting the right optimization algorithm parameters.

Many classical optimization algorithms are not applicable to large designs with many nonlinear constraints. Also, many algorithmic methods, such as gradient-based iterative improvement algorithms, have the problem of getting struck in a local optimum. To obtain a solution close to the global optimum, these methods need a fairly good initial starting point. The generation of this starting point can be performed by means of heuristics, but this suffers from the same limitations as all knowledge-based techniques: it is difficult to formulate in a general way, which holds for any context. In OPASYN [KOH_89, KOH_90], a steepest-descent algorithm is used. However, to overcome the problem of local optima, first a coarse-grid sampling through the domain of the design variables is carried out, and the steepest-descent algorithm is then started from each of these multiple starting points. In addition, the steepest-descent algorithm requires a continuous first derivative and has only limited constraint handling capabilities [KOH_90].

The alternative approach adopted in the OPTIMAN program is the selection of a global and robust optimization algorithm. An obvious choice is then the **simulated annealing algorithm** [KIR_83], sometimes also called probabilistic hill-climbing, which has become very popular in recent years. It is a statistical method to search the global optimum of a function, though at the cost of a large number of function evaluations. It provides general applicability, flexibility and a good quality of the obtained optimum at the expense of considerable CPU times. However, it has to be mentioned that in the ASAIC system any other global optimization algorithm can be used as well. The OPTIMAN program with the simulated annealing algorithm is just one possible, but viable implementation of a general analog design optimization program. Also, if DONALD [SWI_90] manages to generate a good initial starting point, then a steepest-descent optimization algorithm can be used as well.

Summarizing, the sizing of an analog circuit from performance specifications and technology data is in the OPTIMAN program formulated as a constrained optimization problem, based on an analytic description of the circuit behavior and simulated annealing. The **main features of the OPTIMAN program** are :
• it is independent of any circuit topology and technology. All circuit-specific information is combined in the analytic model, in which additional heuristic knowledge about the circuit can always be stored as well. All technology data are read in from a separate technology file, corresponding to the selected process.
• it is general and robust : the program is able to optimize any circuit that is described by an analytic model. This will be illustrated by the examples of section 6.4. Moreover, the use of simulated annealing guarantees to find the global optimum in moderate CPU times, also for circuits of practical size.
• it is very flexible in the definition of the objective function and the constraints: OPTIMAN enables the user to optimize and explore a circuit in all dimensions of the analog design space.
The OPTIMAN program can be used as a stand-alone program to aid circuit designers, or as part of the ASAIC automated analog design system.

6.3. The analog design formulation in OPTIMAN

In OPTIMAN, analog design is formulated as a constrained optimization problem, based on analytic circuit models. This formulation is discussed in more detail in this section. The first subsection 6.3.1 describes the analytic models. Subsection 6.3.2 discusses the optimization variables. The objective function and the constraints are then described in subsection 6.3.3. Finally, subsection 6.3.4 presents the optimization loop and the simulated annealing routine.

6.3.1. The analytic circuit models

The optimization of a circuit in OPTIMAN is based on an analytic model which describes the behavior of this circuit. In general, the analytic model contains :
- a description of the circuit topology
- a summary of the independent design variables
- relationships between dependent and independent design variables
- design equations relating the circuit performance to the design variables
- additional analog expert knowledge about the circuit and other design constraints.

The design equations contain small-signal characteristics, large-signal characteristics and time-domain characteristics. The small-signal characteristics are automatically derived in symbolic form by means of the symbolic simulator ISAAC [GIE_89c]. The other characteristics still have to be supplied by the designer, in a similar way as in IDAC [DEG_87c], OASYS [HAR_89b] and OPASYN [KOH_90]. However, the small-signal characteristics typically cover the largest part of the analytic model for building blocks such as operational amplifiers and filters. Note also that it is always possible to add additional expert knowledge about the circuit to the analytic model under the form of extra analytic equations.

The complete analytic model for a CMOS two-stage Miller-compensated amplifier is given in appendix A. More examples can be found in [GIE_90d]. Analytic models however cannot only be developed for operational amplifiers, but for any type of analog functional block, such as voltage and current references, comparators, output buffers, filters, etc. Analytic models for switched-capacitor filters developed by means of ISAAC can be found in [GIE_90b] and [WAL_91] for instance.

6.3.2. The optimization variables

As design variables all node voltages, bias currents, element values and device sizes in the circuit are taken. All these variables have to be determined and assigned values to in order to fully characterize the design. From these variables all performance characteristics of the circuit can then be calculated. However, in general, these design variables are not totally independent of each other. They are automatically reduced to an independent set, by means of general constraints (such as Kirchhoff's laws), circuit-specific constraints (such as matching information) and designer constraints (such as offset-reduction rules). This set must be as small as possible, the number of *independent variables* being equal to the number of degrees of freedom, and must contain only variables which are independent of each other. If **these independent variables are assigned values to, then the circuit is fully characterized, and all other variables and the total circuit performance can be calculated. Hence, it is quite obvious to use this set of independent variables as optimization variables, which are varied during the optimization in order to**

obtain an optimal circuit performance. This set of independent variables in general is not unique, and the problem, which is currently being solved by the DONALD program [SWI_90], is then the selection of an appropriate set of independent variables for a particular application.

For the design of operational amplifiers, a **useful set of design variables** are the variables which determine the DC operating point of the circuit: bias currents, external bias voltages and transistor saturation voltages. In addition, all values of capacitors and resistors in the schematic have to be included, as well as the width or the length for each MOS transistor. Each MOS transistor requires three variables: its current, its saturation voltage and its width or length. For the CMOS folded-cascode OTA of Fig. 6.3 for example, one then obtains 32 variables (neglecting the bias block M7–M9): I_{1A}, I_{1B}, I_{2A}, I_{2B}, I_{3A}, I_{3B}, I_{4A}, I_{4B}, I_5, I_6, $(V_{GS}-V_T)_{1A}$, $(V_{GS}-V_T)_{1B}$, $(V_{GS}-V_T)_{2A}$, $(V_{GS}-V_T)_{2B}$, $(V_{GS}-V_T)_{3A}$, $(V_{GS}-V_T)_{3B}$, $(V_{GS}-V_T)_{4A}$, $(V_{GS}-V_T)_{4B}$, $(V_{GS}-V_T)_5$, $(V_{GS}-V_T)_6$, L_{1A}, L_{1B}, L_{2A}, L_{2B}, L_{3A}, L_{3B}, L_{4A}, L_{4B}, L_5, L_6, V_{B1} and V_{B2}. Note that in general each transistor requires 3 variables: either the current and the saturation voltage, or the current and the aspect ratio, or the saturation voltage and the aspect ratio, and either the width or the length.

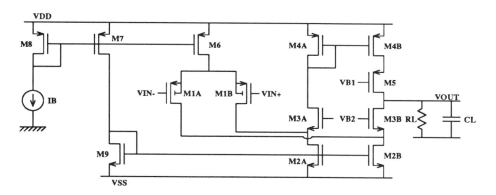

Fig. 6.3. CMOS folded-cascode OTA.

The above 32 variables are not totally independent of each other. They are **reduced to a minimal set of independent variables** by means of :
- general constraints, such as Kirchhoff's current and voltage law
  e.g. $I_{1B} + I_{3A} = I_{2A}$
- circuit-specific constraints, such as matching information
  e.g. $(V_{GS}-V_T)_{4A} = (V_{GS}-V_T)_{4B} = (V_{GS}-V_T)_4$
- designer constraints, such as offset-reduction rules
  e.g. $I_6/2 = I_{4A} = I_{4B}$

By applying these constraints to the CMOS folded-cascode OTA of Fig. 6.3, the 32 design variables can be reduced to 15 independent variables, which are used as optimization variables during the optimization. A possible set of independent variables is then : $\{I_1, (V_{GS}-V_T)_1, (V_{GS}-V_T)_2, (V_{GS}-V_T)_3, (V_{GS}-V_T)_4, (V_{GS}-V_T)_5, (V_{GS}-V_T)_6, V_{B1}, V_{B2}, L_1, L_2, L_3, L_4, L_5, L_6\}$. Note also that the above constraints are all written as analytic equations, which are added to the analytic model of the circuit. If no designer constraints are used, then all equations in the analytic model are inherent to the topology of the circuit under design and independent of the design specifications or design targets. Hence, this analytic model is called an **analytic topology model**. To reduce the degrees of freedom in the design and thus the number of independent variables, the designer can add additional design constraints, based on previous experiences, rules of thumb, heuristics, etc. For example, the currents through the first stage and the second stage of the CMOS folded-cascode OTA of Fig. 6.3 can be selected independently of each other. However, to reduce the offset voltage, the designer can prefer to choose both currents equal to each other, thereby decreasing the degrees of freedom with one. Such designer constraints are often specific for the design targets (e.g. noise minimization) and are thus not generally applicable, even for the same circuit topology. Hence, the resulting analytic model is called an **analytic design model**. It is this model (or the topology model if no designer constraints are added) which is used during the optimization.

The advantage of using transistor saturation voltages instead of aspect ratios as variables is that the operation region of the transistors can easily be controlled by the optimization routine, without extra checking. For example, a MOS transistor is in saturation as long as $V_{DS} \geq (V_{GS}-V_T)$ and is in strong inversion as long as $(V_{GS}-V_T) \geq 0.2$ V. Also, since the optimization variables determine the DC operating point of the circuit, all transistor dimensions W and L and all other transistor parameters, such as g_m and C_{DB}, can be calculated from this DC operating point by means of a general transistor model. Hence, the optimization variables determine the total circuit performance. The transistor model itself may be a first-order model or a more sophisticated SPICE-like one, and is stored as a separate program module. For example, the gain-bandwidth is determined by g_{m1}. However, the relationship between g_{m1} and the optimization variables I_1 and $(V_{GS}-V_T)_1$ is part of the transistor model. The optimization routine in OPTIMAN operates independently of this transistor model. At the same time, all process-dependent information, such as the k_p of a MOS transistor, is stored in a separate technology file. If the process is changed, only a new technology file has to be selected.

6.3.3. The objective function

For analog integrated circuits, the design space generally consists of **four fundamental axes,** as shown in Fig. 6.4 : power, area, speed (frequency range) and dynamic range (limited by noise and distortion). Each particular design can be located in this design space. For most designs, these four characteristics (or part of them) are then optimized. Also, for each design, trade-offs exist between the different axes. For example for a Σ-Δ A/D converter, a trade-off exists between resolution (the signal-to-noise-and-total-harmonic-distortion ratio) and the maximum signal frequency. Hence, it is quite obvious to include these fundamental characteristics in the objective function, with some user-controllable weighting coefficients. In addition, many other circuit characteristics exist as well, which are normally stated with some lower or upper limit. For example, stability considerations can require a minimum phase margin of 45° or 60°.

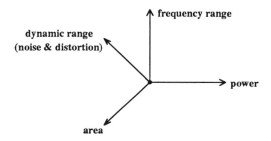

Fig. 6.4. Four fundamental design dimensions for analog integrated circuits.

Therefore, in OPTIMAN, the **performance specifications are treated in two different ways, either as objective or as constraint :**
- performance objectives are absolutely minimized (or maximized). They are incorporated in the objective function. For opamp designs, this function can for instance be the power consumption, the chip area, the noise, the frequency range (bandwidth) or any weighted combination of these four basic design dimensions. The weighting coefficients are then supplied by the user. For example, for an opamp, a mixed power-area minimization can be performed or the trade-off between noise and area can be investigated.
- other performance parameters are considered as constraints (boundary conditions) which have to be satisfied. These constraints can be of inequality type (a specified lower limit or upper limit), or of equality type (a specified center value). For example, for an operational amplifier, a minimum DC gain of 80 dB can be required, whereas for a wideband amplifier a gain of exactly 6 dB can be requested. Also, a systematic offset voltage $V_{os,syst}$. between -1 mV and +1 mV can be demanded.

In addition, other boundary conditions can be imposed on the independent variables or in general on all design variables. A typical example is the requirement for all capacitors or resistors in the circuit to be larger than or equal to zero in order to allow any physical realization. These conditions are usually fulfilled by construction for the independent variables, but have to be checked for all dependent variables as well.

Analog design is thus considered as a **constrained optimization problem** in OPTIMAN. Mathematically, this can be formulated as :

minimize $P(\underline{x})$

subjected to the following constraints :

$$g_i(\underline{x}) \geq 0 \qquad \forall i=1..m$$
$$h_j(\underline{x}) = 0 \qquad \forall j=1..n \qquad\qquad (6.1)$$

where $\underline{x}$ is the vector of the independent variables, $P(\underline{x})$ the objective function, $g_i(\underline{x})$ an inequality constraint and $h_j(\underline{x})$ an equality constraint.

The **objective function** P, which is minimized during the optimization, then consists of a sum of weighted contributions of the different design objectives f_i :

$$P = \sum_i \text{sign}_i \; \alpha_i \; \left(\frac{f_i}{f_i^*}\right) \qquad\qquad (6.2)$$

where sign_i is +1 or -1 depending on whether the objective f_i has to be minimized or maximized, α_i is the weighting coefficient and f_i^* an acceptable value for the objective f_i. The meaning of acceptable value is that the designer is as satisfied with each individual design objective f_i, when it achieves its acceptable value f_i^*. This is similar to the concept of good and bad values in DELIGHT.SPICE [NYE_88]. Mathematically, the acceptable values are used to scale the different contributions in the objective function. Both the weighting coefficients and the acceptable values have to be supplied by the OPTIMAN user. The weighting coefficients then allow him to explore trade-offs between the different design objectives.

The **constraints** can be included in the optimization in two different ways. A first possibility is the inclusion of barrier functions in the objective function. These barrier functions (e.g. exponentials) then penalize constraint violations. A second possibility when using simulated annealing is simply the rejection of a newly generated state which does not satisfy all constraints. In both cases, the constraints can be active from the beginning of the optimization, or they can become more and more active with decreasing temperature towards the end of the annealing process. Barrier functions turn out to be appropriate for descent algorithms, since they always drive the optimization into the valid region. Due to the statistical nature of the simulated annealing algorithm, however, many states outside the valid region are tried out and each time the complete objective function has to be calculated. Experiments have shown that the use of barrier functions requires more CPU time with simulated annealing than simply evaluating the constraints and boundary conditions for each newly generated state and rejecting this state as soon as one constraint is violated. Hence, this rejection technique is adopted in the OPTIMAN program for inequality constraints. A state which satisfies all inequality constraints and boundary conditions is called a feasible state, otherwise it is called an unfeasible state. For equality constraints, on the other hand, the statistical nature of simulated annealing makes the probability of generating a state, which satisfies all equality conditions, practically zero. The largest part of the generated states would be rejected and the number of accepted states would be far too low for finding any acceptable optimum. Therefore, equality constraints $h_j(\underline{x})$ are added to the objective function $P(\underline{x})$ as additional terms which become more and more important with decreasing temperature :

$$P(\underline{x}) = \sum_i \text{sign}_i \ \alpha_i \ (\frac{f_i(\underline{x})}{f_i^*}) \ + \ \sum_j \beta_j \ \frac{|h_j(\underline{x})|}{T} \qquad (6.3)$$

where T is the temperature during the annealing process and β_j is a weighting coefficient. So, in the beginning of the annealing process, the state space is scanned while nearly neglecting the equality conditions. At the end of the annealing process, the terms representing the equalities are minimized to zero and the final solution satisfies the equality constraints.

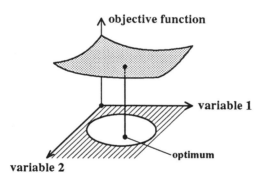

Fig. 6.5. Valid design space for an hypothetical example with two independent variables, and indication of the optimal design point.

It can be concluded from the above discussion that an analog circuit in the OPTIMAN program is regarded as a **multidimensional space**, in which the independent variables are the dimensions. Each combination of values for the independent variables (or each point in the multidimensional space) defines a different instance of the circuit, each with its own performance. The valid design space for a particular application consists of those points which satisfy the design constraints. For the two-dimensional case, this is schematically depicted in Fig. 6.5. To each point in the design space also corresponds a value of the objective function. OPTIMAN then searches in the valid or feasible design space of the independent variables (as determined by the physical boundary conditions and the inequality constraints) for the point, which optimizes the objective function (consisting of the design objectives and equality constraints). This optimum design point is returned by the program.

6.3.4. The optimization loop

The optimization algorithm used in the OPTIMAN program is **simulated annealing** [KIR_83]. This is a general and robust global-optimization method, based on random move generation and statistical move acceptance. Starting from the present state of the optimization problem, the simulated annealing algorithm randomly generates a move to a new state in the state space. Moves which lower the objective function are always accepted and the new state becomes the new starting point. Moves which increase the objective function are accepted or rejected based on a statistical criterion, which is controlled by a parameter. This parameter is called temperature because of the similarities of the simulated annealing algorithm with the physical annealing of materials. The temperature and the acceptance probability of up-hill moves are high in the beginning of the optimization, but are decreased gradually during the course of the optimization. At each temperature, several moves are generated until equilibrium is reached at this temperature. The temperature is then decreased and a new equilibrium is sought. This process is repeated until convergence occurs at lower temperatures.

Simulated annealing is a general and robust optimization method, which does not depend on any rules nor on any mathematical properties of the objective function and its derivatives. Hence, it can be applied to a wide range of different optimization problems, also to problems with many optimization variables such as VLSI placement problems [SEC_85]. As it also allows up-hill moves, the probability of getting trapped in a local optimum is reduced and a solution close to the global optimum can be found at the expense of a large number of function evaluations (due to the statistical nature of the algorithm). Hence, **the simulated annealing algorithm provides general applicability, flexibility and a good quality of the obtained optimum at the expense of considerable CPU times**. The annealing routine used in OPTIMAN is SAMURAI [CAT_88], an efficient kernel with fully adaptive temperature scaling and move range limitation and a novel inner-loop criterion. These features keep the CPU-time consumption still acceptable for large circuits.

To apply simulated annealing to a continuous problem such as analog design optimization, the optimization variables (the independent variables) are gridded over a certain range. A move is then defined as changing the value of one optimization variable with a certain amount (determined statistically by the annealing routine within the move range). The **general optimization loop** is then as follows : the program statistically selects new values for the independent variables in a move range around the present values. It calculates all dependent variables, checks if the corresponding design satisfies all boundary conditions and inequality constraints, calculates the objective function and statistically accepts or rejects the new state. This loop is executed until convergence occurs at lower temperatures, while the move range is gradually decreased with decreasing temperature. In pseudo-code, this can be denoted as :

```
analog_design_optimization_by_simulated_annealing()
{
generate initial solution and initial temperature;
WHILE (temperature_stop_criterion not satisfied) DO
    WHILE (inner_loop_criterion not satisfied) DO
        {generate new values for independent variables;
        check all boundary conditions and constraints;
        calculate objective function;
        IF (accepted with statistical criterion) THEN update solution;
        decrease temperature};
```

At the **initialization**, the independent variables are gridded over their initial range. Simple initial boundaries for the variable values can be derived from a graph representation of the relationships between specifications and variables and from the numerical specification values [GIE_90d]. The initial temperature is determined by randomly exploring the design space with a large number of moves and calculating the temperature value required for a given acceptance probability of say 85% (without equality constraints in the objective function). As initial solution, the first feasible state, which is randomly generated during this initialization phase, is taken.

6.4. Practical design examples

The OPTIMAN program is written in VAX PASCAL and runs on VAX under VMS. The current topology database contains models for several frequently used CMOS operational amplifiers as well as some other analog blocks. The current technology process is the MIETEC 3 μm CMOS n-well process, but other processes can easily be included as well.

The feasibility and general applicability of the OPTIMAN program are now illustrated by means of several design examples [GIE_89a, GIE_90b]. The obtained results are verified by means of numerical simulations and measurements on integrated samples. At the same time, OPTIMAN's performance and efficiency are compared to other programs from the literature.

6.4.1. Example 1: CMOS folded-cascode OTA

The first example is the CMOS folded-cascode OTA of Fig. 6.3. The analytic model of this opamp can be found in [GIE_90d]. A sample has been designed towards minimum power consumption for the specifications given in Table 6.1 below and for a capacitive load of 50 pF with +2.5/-2.5 V supply voltages. The resulting biasing current is 14 μA. The same circuit has then been laid out and processed. Fig. 6.6 shows the die photograph. The design specifications are compared to OPTIMAN data, SPICE simulations and measurement results in Table 6.1. Notice the close correspondence between all values, which demonstrates that simplified analytic models indeed can accurately model the circuit behavior.

Fig. 6.6. Die photograph of the realized CMOS folded-cascode opamp.

By selecting different weighting coefficients in the objective function, the designer can trade off performance specifications against each other (e.g. power versus noise). In subsection 6.3.2, the number of independent variables for the CMOS folded-cascode opamp of Fig. 6.3 has been reduced to 9 (if the 6 independent transistor lengths are not considered). Table 6.2 then compares the final optimized values of these independent variables for three different designs with the same set of specifications already shown in Table 6.1, but with different optimization targets: power minimization, noise minimization, gain-bandwidth maximization and area minimization. All these designs require only a few minutes on a VAX 750

(0.6 Mips). Note that in these optimizations, the saturation voltages are allowed to vary between 0.2 V and 0.5 V.

specification	spec	OPTIMAN	SPICE	measurement
GBW [kHz]	≥ 200	223	224	220
power [mW]	≤ 1	0.140	0.142	0.15
gain [dB]	≥ 60	83	97	86
phase margin [°]	≥ 60	89	86	83
slew rate [V/µs]	≥ 0.13	0.28	0.28	0.22
noise [µV_{RMS}]	≤ 20	16.1	17.6	17.0
output range [V]	≥ 1	1.0	1.07	1.12
input range [V]	≥ 1	1.0	1.09	1.14
offset [mV]	≤ 0.5	0.14	0.1	0.2

Table 6.1. Comparison of specifications, OPTIMAN data, SPICE simulations and measurement results for a minimum power design of the CMOS folded-cascode OTA.

variable	power	noise	GBW	area
I_1	7 µA	49 µA	49 µA	7 µA
$(V_{GS}-V_T)_1$	0.20 V	0.20 V	0.20 V	0.22 V
$(V_{GS}-V_T)_2$	0.41 V	0.50 V	0.22 V	0.50 V
$(V_{GS}-V_T)_3$	0.44 V	0.20 V	0.50 V	0.50 V
$(V_{GS}-V_T)_4$	0.38 V	0.50 V	0.23 V	0.40 V
$(V_{GS}-V_T)_5$	0.29 V	0.20 V	0.50 V	0.36 V
$(V_{GS}-V_T)_6$	0.39 V	0.24 V	0.28 V	0.38 V
V_{B1}	0.15 V	0.15 V	0.10 V	0.10 V
V_{B2}	-0.45 V	-0.60 V	-0.75 V	-0.20 V

Table 6.2. Optimization variables for power minimization, noise minimization, GBW maximization and area minimization of the CMOS folded-cascode OTA.

6.4.2. Example 2: switched-capacitor biquad of Fleischer and Laker

The second example is the switch-noise minimization of the switched-capacitor biquad of Fleischer and Laker [FLE_79], shown in Fig. 6.7. The analytic description can be found in [WAL_87, WAL_91]. The objective is to implement a given filter transfer function, but the capacitor values have to be selected such that the output noise, caused by the switches and integrated over the frequency band of interest, is minimized. Additional constraints are a maximum allowed capacitor area of 100 mm^2 and the scaling of the dynamic range at the outputs of both operational amplifiers in the biquad. The latter is an equality constraint, which is incorporated in the objective function but which becomes important at lower temperature values only.

The design variables are the 12 capacitor values. Since a given biquadratic transfer function with 5 independent coefficients has to be realized, these capacitors are not totally independent of each other and the optimization variables consist of a subset of 7 out the 12 capacitor values. For the OPTIMAN run, the capacitors A, B, D, F, G, I and L have been selected as independent variables. The other capacitors are calculated out of these independent variables by means of the transfer function coefficients. OPTIMAN then determines the 7 capacitor values by applying simulated annealing on the analytic model, thereby reducing the integrated switch noise at the output of the biquad. The same optimization problem has also been solved by the dedicated optimization program BISON [WAL_87, WAL_91], which exploits designer rules to eliminate the degrees of freedom in this biquad. By carefully investigating the objective function and by applying designer knowledge, additional constraints can be provided and the number of independent variables is reduced to 1 for this biquad. This fundamental design variable is then optimized in BISON.

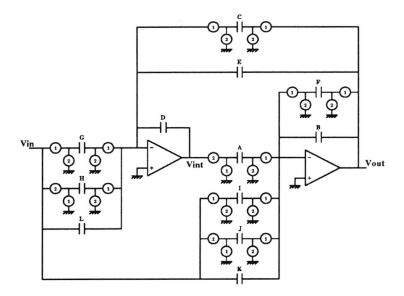

Fig. 6.7. Switched-capacitor biquad of Fleischer and Laker.

Table 6.3 compares the resulting capacitor values from OPTIMAN [GIE_90b] and BISON [WAL_87] a particular audio-band notch filter. All capacitor values from OPTIMAN and BISON, as well as the final output noise, are quite close to each other. A typical example is capacitor J, for which a rule in BISON states that it has to be zero, whereas OPTIMAN obtains a small value. The small differences either indicate that the optimum is not very sensitive to changes in the capacitor values, or that several optima exist with approximately the same objective function value (also called a degenerated optimum). Note also that the capacitor values in Table 6.3 are numerical optimization results (with a grid of 1 fF) and still have to be rounded off for any physical capacitor implementation.

capacitor	OPTIMAN	BISON
A	5.329	5.493
B	13.174	12.250
C	16.489	18.090
D	19.093	20.420
E	14.929	10.130
F	2.015	3.812
G	15.115	16.420
H	0.150	0
I	1.297	1.654
J	0.019	0
K	11.371	11.740
L	0.877	0
noise	19.49 μV	19.43 μV

Table 6.3. Comparison of capacitor values (in pF) for a switch noise minimization of the Laker and Fleischer biquad by OPTIMAN and BISON.

Although the results of Table 6.3 are in good agreement, both programs OPTIMAN [GIE_90b] and BISON [WAL_87] are totally different in concept. BISON obtains the results in a shorter time, a few seconds, but is dedicated to the switch-noise optimization of this single biquad topology only and relies on expert rules. OPTIMAN on the other hand requires more CPU time, a few minutes, but is general and performs a global optimization. This is the well-known trade-off between generality and flexibility on one hand and CPU time on the other hand. Anyway, the above results demonstrate the usefulness and reliability of the OPTIMAN program, but they also verify the rules used in BISON.

6.4.3. Comparison to other approaches

The OPTIMAN program can now be compared to other sizing or dimensioning programs presented in the literature.

OPASYN [KOH_90] somewhat resembles the OPTIMAN approach, since it also performs a weighted optimization based on analytic circuit models. OPASYN, however, uses a steepest-descent optimization algorithm initiated from multiple starting points. In addition, it is limited to CMOS operational amplifiers only and the analytic models are derived completely manually by an expert designer (instead of being generated by a symbolic simulator). The optimization results for the opamps in the program's database, however, are quite good and are obtained quite fast (typically 1 minute of CPU time on a VAX 8800).

COARSE [HEI_89] optimizes CMOS opamps in two phases. First, an appropriate DC operating point is selected based on circuit-specific first-order formulas. Then, the circuit performance is optimized around this operating point with a weighted min-max optimization algorithm. A numerical simulator is used in the optimization loop to evaluate the circuit performance. This approach has been applied to some CMOS opamps, but is difficult to generalize to bipolar opamps and to other circuit classes.

OAC [ONO_90] sizes CMOS opamps in tow steps. In the global design phase, rough device sizing is performed by selecting and modifying an existing design example based on built-in design knowledge. In the detailed design phase, the circuit is further optimized. During each iteration, the layout is generated procedurally and the extracted circuit is simulated and modified according to the nonlinear optimization routine. Because of these iterations over the layout, the CPU times are of course longer, but the quality of the result is quite good as all layout parasitics are fully included in the optimization. An open question still is whether this approach can be extended to other analog blocks as well.

Three other programs, BLADES [ELT_89], IDAC [DEG_87c] and OASYS [HAR_89b], also automatically size analog circuits. These programs are based on knowledge-based techniques and thus produce results in very short CPU times, typically a few seconds. BLADES [ELT_89] and OASYS [HAR_89b], however, lack any real optimization and only produce first-cut designs by executing design rules or design plans, respectively. IDAC [DEG_87c], on the other hand, first produces an initial solution by executing the corresponding formal procedure of the circuit. This initial solution can then be improved with a descent-like optimization algorithm. The problem with these knowledge-based methods, however, is that it is difficult to acquire and formalize the expert knowledge in a context-independent way. As a result, most of the design knowledge is oriented towards some implicit optimization goal. For example, IDAC [DEG_87c] is directed towards low-power low-frequency applications. For other optimization targets, the knowledge-based programs produce results, if any, which can be far from optimal. Hence, knowledge-based programs are perhaps faster than optimization-based programs, such as OPTIMAN [GIE_90b] and OPASYN [KOH_90], but they are less general and less flexible, and the quality of the resulting design still has to be improved with some optimization method. For an automated analog design system, it is not important whether the design is completed in a few seconds or in a few minutes. The most important concern is the quality of the resulting design.

Remark

A final remark about sizing analog circuits concerns the problem of technological variations. Because of these statistical variations and because of temperature changes, the performance of the analog circuit may differ from the nominal behavior. The sizing program has to take these variations into account in order to guarantee that the circuit satisfies the specifications under all conditions. This then requires techniques such as statistical analysis (for example with OASYS [CAR_90]) and design centering (to maximize the yield). The discussion of these techniques however is beyond the scope of this book. It is only worthwhile mentioning here that in the ASAIC system, DONALD [SWI_90] automatically selects the appropriate parameter values to obtain a worst-case result during the optimization with OPTIMAN [GIE_90b].

6.5. Automated layout generation of analog integrated circuits

The final task in the design of analog integrated circuits, after the selection of an appropriate schematic and the sizing of this schematic for the given specifications and technology data, is the generation of the layout. For the sake of completeness, this automated analog layout generation is discussed briefly in this section to indicate the major issues and to point out the major existing approaches. As shown in Fig. 6.8, layout generation is the translation of the sized schematic (structure) into geometry, i.e. determining the shape, place and interconnections of all components in the circuit in the form of rectangles and polygons on several mask layers, in such a way that these masks can be used to fabricate the circuit in silicon. This layout generation process is subjected to several conditions. First of all, the resulting silicon must be guaranteed to be functional within all specifications. The second condition is then to use the chip area as efficiently as possible. Note that the above order of the conditions is important: **for the layout of analog integrated circuits, the functionality prevails above the compactness.** As the performance of analog circuits is very sensitive to layout parasitics, a less compact layout is thus preferred over a nonfunctional circuit.

As for digital circuits, several analog layout styles exist as well with different degrees of customization [ALL_86a], ranging from predefined layouts over semi-custom (analog arrays, standard cells, parameterized procedural generators) to full-custom layouts. These approaches mainly differ in chip density, chip performance, designer flexibility and turnaround time, usually trading off one characteristic against another. However, for analog integrated circuits and especially for ASICs, the flexibility and the achievable performance are the most important factors, since many different circuit schematics are used and the performance specifications and thus the optimal device sizes can widely vary, even for the same schematic. Semi-custom methods based on predefined and/or partially parameterized layouts lack this large flexibility and can result in really poor layouts for certain applications. In addition, design systems based on standard cells or procedural module generators would require the development of a large number of different cells or generators, at least one for each schematic in the system's library. These libraries also have to be updated whenever the technology process changes. This is an enormous effort and cost.

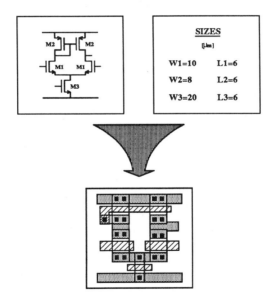

Fig. 6.8. Definition of layout generation. Illustration for an operational amplifier.

In chapter 2, the ASAIC system has been presented for the automated design of analog functional modules starting from module specifications and technology data down to layout [GIE_90a]. According to the ASAIC methodology, the sizing of the analog modules is tailored to each application, in order to obtain the most economical solution with the best overall performance at the lowest cost (in power and area) [GIE_90b]. It is quite obvious that the module then also has to be laid out in a custom way for each application, corresponding to the determined optimal device sizes. Otherwise, all advantages of the optimization are lost. In addition, for many high-performance applications, the layout is critical to ever meet the stringent performance specifications. For these reasons, analog semi-custom layout styles are excluded and the layout is generated from scratch for every application in ASAIC and in most other analog design systems.

The first subsection 6.5.1 now briefly discusses a general strategy for the automated generation of custom layouts for analog integrated circuits. This strategy consists of four steps: generation of structural entities, placement, routing and compaction. Subsection 6.5.2 then describes the major analog layout programs presented in the literature. Finally, subsection 6.5.3 briefly points out the important issues and present results in analog placement and routing.

6.5.1. Strategy for automated analog layout generation

An automated analog module layout generator, which generates a custom layout of a module for each application, takes as input the topological and size information of the module and process and layout parameters. It then generates a fully functional layout of the circuit, which has to be as compact as possible while satisfying all analog layout requirements, such as matching and symmetry conditions, crosstalk avoidance, etc. Because of these analog layout constraints, which are much more complicated and diverse than for digital circuits, existing digital layout tools (e.g. routers) cannot be used for analog applications without substantial modifications. The analog layout program, if intended to be general, must also be independent of any particular module topology and technology. In addition, if possible, it must also allow analog expert layouters to modify the automatically generated layout, if they are not totally satisfied with it. These interactive modifications must be at the symbolic level, such that the correctness of the final layout is guaranteed during all manual operations. This interactive feature will greatly improve the acceptability of an automated analog layout tool in the analog design community. Finally, for higher-level modules, the layout has to be built up hierarchically to speed up the layout generation process.

In general, a topology-independent **strategy for the automated customized layout of analog integrated circuits** then **consists of the following four steps** :
- selection of structural entities out of the circuit schematic
- optimal placement and generation of the entities
- interconnection of all entities with an analog router
- analog compaction of the resulting layout

This strategy then corresponds to a **macro-cell layout style** since it combines component-level procedural layout generators (for the structural entities) with macro-cell-like placement, interconnection and compaction of these entities at the circuit level.

The primitives to be handled by the placement and routing procedures are thus what we call **structural entities**. In the most straightforward way, these entities are individual devices. The many matching requirements between matching devices (which have to be placed as close to each other as possible, with the same shape and orientation), however, can easily overconstrain the placement procedure. Therefore, this problem is often anticipated by considering matching devices as one entity to be placed and interconnected as one macro-cell, and the layout of which is generated in a procedural way, guaranteeing all matching constraints by construction. The structural entities are thus generally defined as functional groups of one or more devices, such as a capacitor, a resistor, a single transistor or two or more matching transistors. For each structural entity, a process-independent parameterized module generator has to be developed which procedurally generates a layout for this entity for the actual parameter values (sizes and shape) of the application. These structural entities are then treated as macro-cells during the remainder of the layout generation process: the entities are optimally placed and interconnected, whereby other analog constraints such as the reduction of routing parasitics are included. Note that each structural entity can be laid out in different ways for the same device sizes: each entity has a number of possible physical realizations, called *variants* (for example a transistor in normal rectangular form, in snake form or in comb form), and each variant can have a number of different ratios or *shapes* (for example different number of combs even for the same width and length of the transistor). The actual variant and shape of each entity are only decided during the placement (if not imposed by the user), i.e. the layouts of the different entities are fitted to each other, in order to obtain a dense final layout. Note that an efficient and frequently used datastructure to represent the different variants and shapes of a structural entity for this placement optimization is a bounding curve (or shape function) [OTT_83]. If a final compaction is carried out after the routing, it of course has to preserve all the symmetry and matching in the circuit.

The placement itself can be accomplished by an algorithmic optimization method or by an expert system which contains layout expertise of how to generate a suitable layout. Nowadays, powerful placement optimization algorithms exist, among which simulated annealing [KIR_83] is very popular. An expert-system-based placement on the other hand can be much faster, but suffers from the problem of generality. Expert rules can readily be provided for a particular circuit schematic and for restricted ranges of device sizes. However, they are difficult to formulate in a general context-independent way, valid for a large range of schematics and device sizes. New rules have to be added for each new topology. At this point, rules suffer from the same drawbacks as procedural circuit layout generators. Therefore, it can be concluded that purely algorithmic approaches are more general and flexible, and applicable to a larger category of circuits.

6.5.2. Major automated analog layout approaches from the literature

For the automated generation of custom analog layouts, several approaches and tools have already been published [BER_88b, BOW_89, CHE_89, CHO_90ab, COH_90, GAR_88, GAT_89, GYU_89, KAW_88, KAY_88ab,89, KIM_89, KOH_88,89,90, KUH_87, MAL_90, PIG_89,90, RIJ_88,89, SCH_88, TRO_89a, WIN_87, YAG_86]. The most viable approaches are now discussed in somewhat more detail to reflect the state of the art in analog layout generation. It is however important to mention here that all the approaches presented below use the above macro-cell place-and-route layout style. Apparently, **there seems to be a clear consensus in the analog CAD community with respect to the layout strategy to be followed.** The programs only differ in the implementation details of the different tasks (module generation, placement, routing and compaction) within this layout strategy.

6.5.2.1. ILAC

The ILAC program, developed at CSEM, automatically generates a custom layout for the IDAC-sized circuit in a macro-cell place-and-route style [RIJ_89]. It handles typically analog layout constraints, such as device matching, symmetry, and distance and coupling constraints. The circuit is first partitioned into structural entities, the layout of which is generated by specialized procedural module generators. The structural entities are then optimally placed and routed, and the resulting layout is compacted. The program is fully automatic, but the designer can also modify the layout interactively at the symbolic level if he wants to.

6.5.2.2. KOAN/ANAGRAM

The KOAN/ANAGRAM program, developed at Carnegie Mellon University, generates a custom layout for the OASYS-sized schematic in a macro-cell place-and-route style [GAR_88, COH_90]. The KOHN placement program contains a few device-level module generators. The resulting entity layouts are then optimally placed, while handling symmetry and matching conditions, on-the-fly device mergings, etc. The ANAGRAM analog router then performs a detailed routing, including symmetrical wiring and penalty-driven crosstalk avoidance.

6.5.2.3. OPASYN

The OPASYN program, developed at the University of California, Berkeley, also generates a custom layout of the sized opamp schematic in a macro-cell place-and-route style [KOH_90]. The fixed opamp schematics are a priori divided by the designer in structural entities, which are generated by parameterized leaf-cell generators. The placement is then performed according to a predefined circuit-specific slicing tree (placement template), starting from a user-supplied aspect ratio or size constraint. For the routing, all signal nets are divided in different classes (e.g. sensitive nets, noisy nets...), which are routed subsequently with the rip-up-and-reroute router MIGHTY [SHI_86]. Final compaction then yields the mask-level layout. In CADICS [JUS_90], on the other hand, hierarchical templates are used in which the placement is optimized at lower levels, while the overall floorplan organization is fixed. In addition, more advanced analog routers are used in this program.

6.5.2.4. SALIM

The SALIM program, developed at the Ecole Polytechnique Fédérale de Lausanne, also generates a custom layout in a macro-cell place-and-route style [KAY_88b, KAY_89]. The program provides both an interactive mode for experienced designers and an automatic mode for inexperienced designers. In the automatic mode, either a rule-based expert system can be used, which then controls a set of procedural algorithms as subtools, or a purely algorithmic approach based on simulated annealing can be applied. The designer has to divide the circuit in structural entities, for which specialized layout generators exist. These structural entities are then placed, either interactively, either by the expert system, or by means of simulated annealing (or by a mixture of the latter two). An analog-oriented router [PIG_89, PIG_90] then completes the layout.

6.5.2.5. Comparison

Although the above programs all follow the same layout strategy, a distinction can still be made based on functionality and overall implementation approach. First, with respect to functionality, only ILAC [RIJ_89] and SALIM [KAY_88b] offer an interactive mode in addition to the automatic mode. This interactive mode is useful for experienced designers, who always want to improve the automatically generated layouts themselves, but is of no importance for system designers, who simply want a good layout in a short time without any interaction. Secondly, with respect to the overall implementation approach, SALIM [KAY_88b] is the only program which uses a rule-based approach in combination with procedural algorithms. All other programs use a purely algorithmic approach. The use of expert layout rules can perhaps be a convenient way of dealing with analog layout constraints, but it suffers from the problems of generality and flexibility as mentioned above.

6.5.3. Analog placement and routing

The task of the **analog placement (sometimes also called floorplanning)** is to geometrically place the different structural entities in the circuit on the chip area, such that :
• all analog constraints (matching, symmetry, clustering or separation) are satisfied

- the occupied area is as compact as possible, within some user-supplied aspect ratio constraints
- the length of the interconnections is reduced by placing strongly connected entities close to each other.

This clearly is an optimization problem, which at the same time determines the actual implementation form of each structural entity.

There are mainly two different ways to represent a placement (or floorplan) during this optimization:

- represent each entity by its center coordinates and its orientation. This technique results in larger search spaces and thus CPU times, since the available area is gridded and the entities are placed on absolute coordinates. But it may result in denser layouts.
- represent the whole floorplan by a slicing structure obtained by repeatedly partitioning the floorplan into horizontal and/or vertical slices [WON_86]. Each partition is then assigned to one particular entity. This method only determines the relative placement of the different structural entities and can easily deal with entities with flexible dimensions. This, however, usually results in less dense layouts, which strongly resemble the blocks-and-channels look of digital layouts.

The selection between these two floorplan representation forms comes to the trade-off between CPU time and quality of the resulting floorplan. Slicing structures are used in ILAC [RIJ_89], OPASYN [KOH_90] and SALIM [KAY_88b], whereas KOAN [COH_90] performs a flat floorplan optimization without slicing structures. For the optimization algorithm itself, most programs use simulated annealing which is appropriate for these problems with multiple constraints.

ILAC [RIJ_89] uses simulated annealing and slicing structures, and optimizes the floorplan in two steps. First, an initial floorplan is obtained with rough estimations for the channel sizes. Secondly, the floorplan and the global routing are optimized simultaneously. KOAN [COH_90] on the other hand performs a flat annealing (with absolute cell coordinates). The program has only a limited number of module generators, one for each type of device, and the placement routine takes care of all analog constraints. The placement in OPASYN [KOH_90] is template-driven: the slicing structure for each opamp topology is not determined in an optimization loop, but is predefined. Finally, SALIM [KAY_88ab] offers three placement modes to the user: interactive, automatic by expert system or automatic by simulated annealing based on slicing structures.

After the placement, the structural entities then have to be **interconnected** [OHT_86]. This phase, however, is critical for the overall performance of the circuit, since it introduces most of the layout parasitics. Especially for analog circuits, many constraints are imposed on the router to obtain well-functioning circuits. These include the reduction of routing parasitics, avoidance of parasitic couplings (crosstalk) between different nets, distinction between different net categories (power and ground, sensitive, noncritical and noisy nets) with different routing priority, symmetrical wiring, etc.

With respect to the routing of analog integrated circuits, several approaches have already been published in the literature [CHO_90ab, COH_90, GAR_88, GYU_89, MAL_90, PIG_89,90, RIJ_89], all handling analog constraints to some extent. In ILAC [RIJ_89], the global routing is performed by a best-first maze search, and the subsequent detailed routing by a gridless scan-line-based incremental channel router, minimizing parasitics and couplings. ILAC also allows the designer to interactively modify the routing at the symbolic level. ANAGRAM II [COH_90] performs detailed routing with a gridless line-expansion algorithm, with penalty-driven crosstalk avoidance, geometrically symmetric differential wiring and greedy rip-up-and-reroute of nets. The routing in SALIM [PIG_89] consists of path scheduling and path optimization, followed by greedy-like gridless channel routing. Finally, constraint-driven channel and area routing is presented in [CHO_90ab, GYU_89, MAL_90]. Here, by using sensitivities, limits on circuit performance deviations are translated in maximum boundaries on layout parasitics, which drive the routing. In this way, time-consuming iterations over layout generation and layout extraction are avoided.

6.6. Conclusions

In this chapter, it has been illustrated how the symbolic analysis techniques presented in the previous chapters can be used to size practical circuits and to create an open analog design system. To this end, the program OPTIMAN [GIE_89a, GIE_90b] has been discussed, which optimally sizes a circuit based on the corresponding analytic model. These analytic models, which fully characterize the circuit behavior, can to a large extent be generated automatically by means of a symbolic simulator. Algorithmic details of the program have been provided. The approach has then been illustrated with practical design examples, and has been compared to other analog sizing programs from the literature.

Finally, to complete the design cycle, a short discussion of analog layout generation has been provided. A general strategy for the automated generation of custom layouts for analog circuits has been discussed. This strategy applies a macro-cell place-and-route style and generally consists of structural entity generation, analog placement, routing and compaction. The major issues and approaches in analog placement and routing have been presented briefly.

We hope that the above discussion has provided the reader with a clear idea of the state of the art in the increasingly important field of analog symbolic simulation and analog design automation in general.

A CHARACTERIZATION OF A CMOS TWO-STAGE OPAMP

In this appendix, a fully analytic characterization of a CMOS operational amplifier is presented. First, a fairly complete list of characteristics for any operational amplifier is given. These characteristics are then derived for a CMOS two-stage Miller-compensated opamp. This results in an analytic model, which is used for automatically sizing the opamp in our analog design methodology. Finally, the numerical results for a typical CMOS two-stage opamp design are summarized.

A.1. Full set of characteristics for an operational amplifier

A fairly complete list of characteristics, which are important for an operational amplifier, is given below. The characteristics are divided in three classes: DC characteristics, AC and transient characteristics, and second-order characteristics.

1. DC characteristics
1.1. Common-mode input voltage range versus supply voltage
1.2. Output voltage range versus supply voltage
1.3. Maximum output current (sink and source)
2. AC and transient characteristics
2.1. Differential-mode gain versus frequency
2.2. Gain-bandwidth versus biasing current
2.3. Slew rate versus load capacitance
2.4. Output voltage range versus frequency
2.5. Settling time
2.6. Open-loop input impedance versus frequency
2.7. Closed-loop input impedance versus frequency
2.8. Open-loop output impedance versus frequency
2.9. Closed-loop output impedance versus frequency
3. Second-order characteristics
3.1. CMRR versus frequency
3.2. Offset voltage versus common-mode input voltage
3.3. Drift versus supply voltage, temperature and time
3.4. Input bias current and input offset current
3.5. Equivalent input noise voltage versus frequency
3.6. Equivalent input noise current versus frequency
3.7. PSRR versus frequency
3.8. Harmonic distortion

A.2. Analytic model for a CMOS two-stage opamp

Most of the above characteristics are now presented in analytic form for the CMOS two-stage Miller-compensated opamp of Fig. A.1. This opamp schematic has been selected here because it is a simple circuit which clearly illustrates the idea of an analytic model to the reader, and because of the availability of integrated samples [SAN_88b]. The analytic equations below fully characterize the opamp and can thus be used to automatically design the amplifier. Since all characteristics are inherent to this particular circuit topology, they form the *analytic topology model* of the circuit. For easing the automatic sizing, the designer can add additional constraints or rules (which are often specific for the given design specifications or targets), resulting in the *analytic design model*. OPTIMAN [GIE_89a, GIE_90b] then optimizes the circuit performance based on the resulting analytic model.

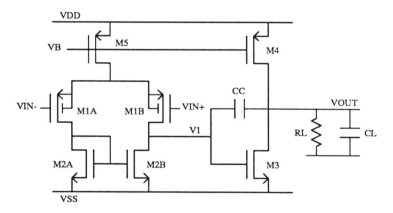

Fig. A.1. CMOS two-stage Miller-compensated opamp.

The characteristics of the CMOS two-stage opamp of Fig. A.1 are now summarized below.

A.2.1. DC characteristics

A.2.1.1. Common-mode input voltage range versus supply voltage

The common-mode input voltage range is limited when some transistor goes out of saturation. For the positive swing, this occurs when M5 goes out of saturation :

$$V_{in,max} = V_{DD} - V_{DSAT5} - V_{GS1} \tag{A.1}$$

For the negative swing, this occurs when M1A goes out of saturation :

$$V_{in,min} = V_{SS} + V_{GS2A} + V_{DSAT1A} - V_{GS1A} \tag{A.2}$$

A.2.1.2. Output voltage range versus supply voltage

The output voltage range is limited when one of the output transistors is driven out of saturation, but it is also limited by the finite current available to drive the load. For the positive swing, this yields :

$$V_{out,max} = min \{V_{DD} - V_{DSAT4} , I_4.R_L\} \tag{A.3}$$

For the negative swing, the output is limited at :

$$V_{out,min} = V_{SS} + V_{DSAT3} \tag{A.4}$$

A.2.1.3. Maximum output current (sink and source)

The maximum source output current is $I_{source} = I_4$, when M3 is cut off. The maximum sink output current is reached when node 1 drives M2B out of saturation :

$$I_{sink} = 0.5 \beta_3 (V_{in+} + V_{GS1B} - V_{DSAT1B} - V_{SS} - V_{T3})^2 - I_4 \tag{A.5}$$

A.2.2. AC and transient characteristics

A.2.2.1. Differential-mode gain versus frequency

The equivalent small-signal circuit of the CMOS two-stage Miller-compensated opamp is shown in Fig. A.2. The expressions for the open-loop gain, the poles and zeros are :

$$A_o = \frac{g_{m1} \, g_{m3}}{(g_{o1B} + g_{o2B}) \, (g_{o3} + g_{o4} + g_L)} \tag{A.6}$$

$$z_1 = + \frac{g_{m3}}{2\pi\ C_C} \tag{A.7}$$

$$z_2 = - \frac{2\ g_{m2A}}{2\pi\ (C_{GS2A} + C_{GS2B} + C_{DB2A} + C_{DB1A})} \tag{A.8}$$

$$p_1 = - \frac{(g_{o1B} + g_{o2B})\ (g_{o3} + g_{o4} + g_L)}{2\pi\ g_{m3}\ C_C} \tag{A.9}$$

$$p_2 = - \frac{g_{m3}\ C_C}{2\pi\ (C_L\ C_C + C_L\ C_1 + C_C\ C_1)} \tag{A.10}$$

where $\quad C_1 = C_{GS3} + C_{DB2B} + C_{DB1B}$

$$p_3 = - \frac{g_{m2A}}{2\pi\ (C_{GS2A} + C_{GS2B} + C_{DB2A} + C_{DB1A})} \tag{A.11}$$

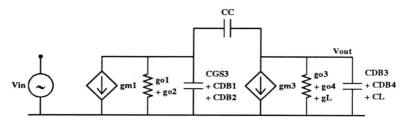

Fig. A.2. Equivalent small-signal circuit of the CMOS two-stage
Miller-compensated opamp.

The gain-bandwidth, the unity-gain frequency and the phase margin are given by :

$$GBW = - \frac{g_{m1}}{2\pi\ C_C} \tag{A.12}$$

$$f_u = GBW\ .\ \sin(PM) \tag{A.13}$$

$$PM = 90° - atan(\frac{f_u}{p_2}) - atan(\frac{f_u}{p_3}) - atan(\frac{f_u}{z_1}) + atan(\frac{f_u}{z_2}) \tag{14}$$

A.2.2.2. Slew rate versus load capacitance

The slew rate is the minimum of the "internal" and the "external" slew rate :

$$SR = \min \left\{ \frac{I_5}{C_C} , \frac{I_4 - I_5}{C_L + C_{DB3} + C_{DB4}} \right\} \tag{A.15}$$

A.2.2.3. Output voltage range versus frequency

The output voltage range is limited by the supply voltage at lower frequencies (see A.2.1.2 above) and by the slew rate at higher frequencies :

$$V_{out} = \frac{SR}{2\pi f} \tag{A.16}$$

A.2.2.4. Settling time

The CMOS two-stage Miller-compensated opamp of Fig. A.1 can be considered as a second-order system. The settling time thus depends on the phase margin of the opamp. If the system is underdamped (i.e. phase margin $< 76.35°$), an oscillating step response is observed and the settling time for a settling margin ε is given by :

$$T_s(\varepsilon) = -\frac{1}{GBW} \frac{2 \cos(PM)}{\sin 2(PM)} \left[\ln(\varepsilon) + \frac{1}{2} \ln\left(1 - \frac{\sin^2(PM)}{4 \cos(PM)}\right) \right] \tag{17}$$

If, on the other hand, the system is overdamped (i.e. phase margin $> 76.35°$), an exponential step response is observed and the settling time for a settling margin ε is given by :

$$T_s(\varepsilon) = \frac{- \ln(\varepsilon)}{GBW \dfrac{\sin^2(PM)}{2 \cos(PM)} \left[1 - \sqrt{\left(1 - \dfrac{4 \cos(PM)}{\sin^2(PM)}\right)} \right]} \tag{A.18}$$

A.2.2.5. Open-loop input impedance versus frequency

The open-loop input impedance for the negative input is :

$$z_{in-} = \frac{1}{s \left[\dfrac{C_{GS1A}}{2} + C_{GD1A} \left(1 + \dfrac{g_{m1A}}{2 g_{m2A}}\right) \right]} \tag{A.19}$$

The open-loop input impedance for the positive input is :

$$Z_{in+} = \cfrac{1}{s\ [\cfrac{C_{GS1B}}{2} + C_{GD1B}\ (1 + \cfrac{g_{m1B}}{g_{o1B} + g_{o2B} + s\ C_C\ M})\ ]} \qquad (A.20)$$

$$\text{with } M = \cfrac{g_{m3}}{g_{o3} + g_{o4} + g_L}$$

A.2.2.6. Closed-loop input impedance versus frequency

The input impedance for unity feedback is :

$$Z_{inc} = \cfrac{1}{s\ [\cfrac{C_{GS1B}}{2}\ (\cfrac{1}{1 + A}) + C_{GD1B}]} \qquad (A.21)$$

$$\text{with } A = \cfrac{g_{m1}\ g_{m3}}{(g_{o1B} + g_{o2B})\ (g_{o3} + g_{o4} + g_L) + g_{m3}\ C_C\ s}$$

A.2.2.7. Open-loop output impedance versus frequency

The open-loop output impedance is given by :

$$Z_{out} = \cfrac{g_{o1B} + g_{o2B} + s\ (C_C + C_1)}{(g_{o1B} + g_{o2B})\ (g_{o3} + g_{o4}) + s\ g_{m3}\ C_C + s^2\ C_C\ (C_1 + C_{DB3} + C_{DB4})} \qquad (A.22)$$

$$\text{where } C_1 = C_{GS3} + C_{DB1B} + C_{DB2B}$$

A.2.2.8. The closed-loop output impedance versus frequency

The output impedance in unity feedback is given by :

$$Z_{outc} = \cfrac{g_{o1B} + g_{o2B} + s\ (C_C + C_1)}{g_{m1}\ g_{m3} + s\ g_{m3}\ C_C + s^2\ C_C\ (C_1 + C_{DB3} + C_{DB4})} \qquad (A.23)$$

$$\text{where } C_1 = C_{GS3} + C_{DB1B} + C_{DB2B}$$

A.2.3. Second-order characteristics

A.2.3.1. CMRR versus frequency

The CMRR of the CMOS two-stage Miller-compensated opamp of Fig. A.1 (with mismatches included) is given by :

$$CMRR = \cfrac{g_{m1}}{\cfrac{(g_{o5} + s\,C_3)}{2}\left[\cfrac{g_{o1} + g_{o2} + s\,C_2}{g_{m2}} + \cfrac{\Delta g_{m1BA}}{g_{m1}} - \cfrac{\Delta g_{m2BA}}{g_{m2}}\right] + g_{o1}\left[\cfrac{\Delta g_{m1BA}}{g_{m1}} - \cfrac{\Delta g_{o1BA}}{g_{o1}}\right]} \tag{A.24}$$

$$\text{where} \quad C_2 = C_{GS2A} + C_{GS2B} + C_{DB2A} + C_{DB1A}$$
$$C_3 = C_{DB5} + C_{n-well}$$

A.2.3.2. Offset voltage

The systematic offset voltage is determined by the difference in voltage at the drain of M2A (M1A) and M2B (M1B), and is given by :

$$V_{os,syst.} = \frac{g_{o1} + g_{o2}}{g_{m1}}\,\Delta V_{DS2BA} \tag{A.25}$$

The statistical offset voltage, on the other hand, is totally determined by mismatches between the transistors M1A-M1B and M2A-M2B.

A.2.3.3. Equivalent input noise voltage versus frequency

The equivalent input noise for frequencies up till the gain-bandwidth can be approximated as :

$$dv_{eq}^2 = 2\,dv_{n1}^2 + 2\,dv_{n2}^2\,(\frac{g_{m2}}{g_{m1}})^2 \tag{A.26}$$

The integrated output noise for unity feedback is then approximately given by :

$$V_{RMS}^2 = \pi/2\,GBW\,dv_{eq,w}^2 + \ln(GBW)\,dv_{eq,f}^2 \tag{A.27}$$

A.2.3.4. PSRR± versus frequency

The PSRR+ with mismatches and V_B shorted to V_{DD} is given by :

$$PSRR^+ = \cfrac{g_{m1}}{-\cfrac{(g_{o5} + s\ C_3)}{2}\ [\cfrac{g_{o1} + g_{o2} + s\ C_3}{g_{m2}} + \cfrac{\Delta g_{m1BA}}{g_{m1}} - \cfrac{\Delta g_{m2BA}}{g_{m2}}] + (g_{o4} + s\ C_{DB4})\ \cfrac{(g_{o1} + g_{o2} + s\ (C_C + C_1)}{g_{m3}}} \tag{A.28}$$

$$\text{where}\quad C_1 = C_{GS3} + C_{DB1B} + C_{DB2B}$$
$$C_2 = C_{GS2A} + C_{GS2B} + C_{DB2A} + C_{DB1A}$$
$$C_3 = C_{DB5} + C_{n-well}$$

A.2.3.5. PSRR- versus frequency

The PSRR- with mismatches and V_B shorted to V_{DD} is given by :

$$PSRR^- = \cfrac{g_{m1}}{(g_{o3} + s\ C_{DB3})\ \cfrac{(g_{o1} + g_{o2} + s\ (C_C + C_1))}{g_{m3}} + g_{o1}\ [\cfrac{\Delta g_{o1BA}}{g_{o1}} - \cfrac{\Delta g_{m1BA}}{g_{m1}}] + s\ C_C} \tag{A.29}$$

$$\text{where}\quad C_1 = C_{GS3} + C_{DB1B} + C_{DB2B}$$

A.2.3.6. Harmonic distortion

The second harmonic distortion ratio at the output of the CMOS two-stage opamp of Fig. A.1 in open loop for a differential-input amplitude A is at lower frequencies given by :

$$HD_2 = -\ A\ \frac{K_2 \cdot g_{m3}}{g_{m3}}\ \frac{g_{m1}}{g_{o1} + g_{o2}} \tag{A.30}$$

In closed loop, this is divided by the square of the loop gain. The third harmonic distortion, on the other hand, is usually much smaller than the second harmonic distortion for CMOS circuits.

A.2.3.7. Influence of temperature on the characteristics

On the condition that the biasing is to first order insensitive to temperature variations, the most important changes come from the decrease of the mobility with increasing temperature. It can then easily be investigated that the DC gain, the gain-bandwidth, the second pole and the Miller zero decrease with increasing temperature, while the first pole increases with increasing temperature. As a result, the phase margin is to first order independent of the temperature.

A.3. Design results for the standard CMOS two-stage opamp

With the above analytic model, a CMOS two-stage Miller-compensated opamp has been designed (see chapter 2) intended for standard applications in the ESAT-MICAS group. Since this opamp is used many times as example circuit throughout the book, the parameter values and design results are presented here briefly.

The design specifications for this standard opamp are :
$\text{GBW} \geq 1 \text{ MHz}$
$C_L = 10 \text{ pF}$
$R_L = 100 \text{ k}\Omega$
$V_{DD}/V_{SS} = \pm 2.5 \text{ V}$
$A_o \geq 1000 = 60 \text{ dB}$
$\text{PM} \geq 60°$
output range $\geq \pm 2 \text{ V}$
systematic offset voltage $< 0.5 \text{ mV}$
minimize power consumption !

The bias current and the compensation capacitor for the resulting design are then given by $I_B = 2.5 \text{ µA}$ and $C_C = 1 \text{ pF}$. All transistor parameter values of the resulting design (as obtained by SPICE [VLA_80b]) are now summarized in the following tables. The circuit schematic is shown in Fig. A.1.

	n/p	W_M	L_M	β	I_D	V_{GS}	V_{DSAT}	V_{DS}
		[µm]	[µm]	[µA/V²]	[µA]	[V]	[V]	[V]
M1A	pMOS	26	16	28	1.25	1.24	0.26	2.59
M1B	pMOS	26	16	28	1.25	1.24	0.26	2.63
M2A	nMOS	10	10	48	1.25	1.14	0.21	1.14
M2B	nMOS	10	10	48	1.25	1.14	0.21	1.11
M3	nMOS	115	5	1252	25.2	1.11	0.18	2.50
M4	pMOS	55	5	214	25.2	1.43	0.42	2.50
M5	pMOS	13	10	22	2.50	1.43	0.42	1.26

	n/p	g_m	g_{mb}	g_o	C_{GS}	C_{GD}	C_{SB}	C_{DB}
		[μS]	[μS]	[μS]	[fF]	[fF]	[fF]	[fF]
M1A	pMOS	7.5	2.1	0.006	216	7	97	50
M1B	pMOS	7.5	2.1	0.006	216	7	97	50
M2A	nMOS	10.3	1.7	0.024	49	2	17	12
M2B	nMOS	10.3	1.7	0.024	49	2	17	12
M3	nMOS	246	41.6	1.18	302	20	124	68
M4	pMOS	95.5	25.3	0.612	141	15	195	101
M5	pMOS	9.5	2.5	0.013	65	3	52	34

The performance characteristics of the designed standard two-stage amplifier are now summarized in the following table. The second column gives the values as calculated with the analytic model. The third column gives the values as obtained with SPICE [VLA_80b] simulations.

parameter	calculated	simulated
DC gain	74 dB	74 dB
gain-bandwidth	1.19 MHz	1.12 MHz
phase margin	67°	66°
slew rate	2.23 V/μs	-
input range	0.84 V	-
output range	2.08 V	-
systematic offset voltage	0.12 mV	0.2 mV
statistic offset voltage	1.7 mV	1.6 mV
integrated output noise	138 μV_{RMS}	-
CMRR @ DC	98.8 dB	98.8 dB
PSRR+ @ DC	110.1 dB	110.5 dB
PSRR- @ DC	96.5 dB	96.4 dB

This concludes our discussion of the CMOS two-stage Miller-compensated opamp. In this appendix, a complete list of performance characteristics of an operational amplifier has been provided. In addition, a detailed analytic model for the CMOS two-stage opamp has been presented, as well as a set of parameters resulting from a practical design, which is frequently used as example circuit throughout the book.

REFERENCES

[ADB_80] P. R. Adby, "Applied circuit theory: matrix and computer methods," Ellis Horwood, pp. 269-343, 1980.

[ALD_73] G. E. Alderson, P. M. Lin, "Computer generation of symbolic network functions - a new theory and implementation," IEEE Transactions on Circuit Theory, Vol. CT-20, No. 1, pp. 48-56, January 1973.

[ALL_83a] P. E. Allen, H. Nevarez-Lozano, "Automated design of MOS op amps," proc. ISCAS, pp. 1286-1289, 1983.

[ALL_83b] P. E. Allen, H. Rafat, E. Macaluso, A. Nedungadi, "AIDE - Analog Integrated circuit Design Enhancement program," proc. Midwest Symposium on Circuits and Systems, 1983.

[ALL_84] P. E. Allen, H. A. Rafat, S. F. Bily, E. Macaluso, A. Nedungadi, "The use of a CAD tool to design analog CMOS integrated circuits," proc. ISCAS, 1984.

[ALL_85] P. E. Allen, E. R. Macaluso, S. F. Bily, A. P. Nedungadi, "AIDE2: an automated analog IC design system," proc. CICC, pp. 498-501, 1985.

[ALL_86a] P. E. Allen, "A tutorial - computer-aided design of analog integrated circuits," proc. CICC, pp. 608-616, 1986.

[ALL_86b] P. E. Allen, P. R. Barton, "A silicon compiler for successive approximation A/D and D/A converters," proc. CICC, pp. 552-555, 1986.

[ALL_87] P. E. Allen, D. R. Holberg, "CMOS analog circuit design," Holt, Rinehart and Winston (New York), 1987.

[ASS_87] J. Assael, P. Senn, M. Tawfik, "A switched capacitor filter silicon compiler," proc. ECCTD, pp. 901-906, 1987.

[ASS_88] J. Assael, P. Senn, M. S. Tawfik, "A switched-capacitor filter silicon compiler," IEEE Journal of Solid-State Circuits, Vol. SC-23, No. 1, pp. 166-174, February 1988.

[BAR_89] A. Barlow, K. Takasuka, Y. Nambu, T. Adachi, J.-I. Konno, "An integrated switched capacitor filter design system," proc. CICC, pp. 4.5.1-5, 1989.

[BER_88a] E. Berkcan, M. d'Abreu, W. Laughton, "Analog compilation based on successive decompositions," proc. Design Automation Conference, pp. 369-375, 1988.

[BER_88b] E. Berkcan, M. d'Abreu, "Physical assembly for analog compilation of high voltage ICs," proc. CICC, pp. 14.3.1-7, 1988.

[BER_89] E. Berkcan, C. K. Kim, B. Currin, M. d'Abreu, "From analog design description to layout: a new approach to analog silicon compilation," proc. CICC, pp. 4.4.1-4, 1989.

[BOL_89a] I. Bolsens, "Electrical correctness verification of MOS digital circuits using expert system and symbolic analysis techniques," PhD dissertation, Katholieke Universiteit Leuven, 1989.

[BOL_89b] I. Bolsens, "Symbolic analysis of MOS VLSI digital networks," proc. BIRA workshop Technical Applications of Automatic Formula Manipulation, Leuven, November 1989.

[BON_80] M. Bon, A. Konczykowska, "A topological analysis program for switched-capacitor networks with symbolic capacitor and switching functions," proc. ECCTD, pp. 159-164, 1980.

[BOS_89] S. Bose, A. L. Fisher, "Verifying pipelined hardware using symbolic logic simulation," proc. ICCD, pp. 217-221, 1989.

[BOW_85] R. J. Bowman, D. J. Lane, "A knowledge-based system for analog integrated circuit design," proc. ICCAD, pp. 210-212, 1985.

[BOW_88] R. J. Bowman, "Analog integrated circuit design conceptualization," presymposium on Expert System Tools for Analog Signal Processing, ISCAS, 1988.

[BOW_89] R. J. Bowman, "An imaging model for analog macrocell layout generation," proc. ISCAS, pp. 1127-1130, 1989.

[BRA_81] R. K. Brayton, G. D. Hachtel, A. L. Sangiovanni-Vincentelli, "A survey of optimization techniques for integrated-circuit design," Proceedings of the IEEE, Vol. 69, No. 10, pp. 1334-1364, October 1981.

[BRA_89] A. Brambilla, G. Cottafava, P. Gubian, "NICE: an optimization system for computer aided design of electrical circuits," proc. ISCAS, pp. 2016-2019, 1989.

[BRU_83] U. W. Brugger, G. S. Moschytz, E. Hökenek, "SFG analysis of SC networks comprising integrators," proc. ISCAS, pp. 68-71, 1983.

[BRY_87] R. Bryant, "Boolean analysis of MOS circuits," IEEE Transactions on Computer-Aided Design, Vol. CAD-6, No. 4, pp. 634-649, July 1987.

[BUS_74] J. J. Bussgang, L. Ehrman, J. W. Graham, "Analysis of nonlinear systems with multiple inputs," Proceedings of the IEEE, Vol. 62, No. 8, pp. 1088-1119, August 1974.

[CAL_89] L. Callewaert, W. Eyckmans, W. Sansen, V. Budihartono, F. Newcomer, R. Van Berg, J. Van der Spiegel, S. Tedja, H. Williams, A. Stevens, "Front end and signal processing electronics for detectors at high luminosity colliders," IEEE Transactions on Nuclear Science, Vol. 36, No. 1, pp. 446-450, February 1989.

[CAM_90] R. Camposano, "High-level synthesis," tutorial High-level Synthesis part I, European Design Automation Conference, 1990.

[CAR_88a] L. R. Carley, "Analog circuit synthesis and exploration in OASYS," presymposium on Expert System Tools for Analog Signal Processing, ISCAS, 1988.

[CAR_88b] L. R. Carley, R. A. Rutenbar, "How to automate analog IC designs," IEEE Spectrum, pp. 26-30, August 1988.

[CAR_88c] L. R. Carley, R. A. Rutenbar, "Automatic synthesis of analog integrated circuits - a tutorial," presymposium ICCAD, 1988.

[CAR_89] L. R. Carley, D. Garrod, R. Harjani, J. Kelly, T. Lim, E. Ochotta, R. A. Rutenbar, "ACACIA: the CMU analog design system," proc. CICC, pp. 4.3.1-5, 1989.

[CAR_90] L. R. Carley, "Automating the design and layout of analog circuits," proc. Educational Sessions of CICC, 1990.

[CAT_88] F. Catthoor, J. Vandewalle, H. De Man, "SAMURAI: a general and efficient simulated-annealing schedule with fully adaptive annealing parameters," Integration, the VLSI Journal (North Holland), No. 6, pp. 147-178, June 1988.

[CEN_80] G. Centkowski, J. Starzyk, E. Sliwa, "Computer implementation of topological method in the analysis of large linear networks," proc. ECCTD, pp. 69-74, 1980.

[CEN_81] G. Centkowski, J. Starzyk, E. Sliwa, "Symbolic analysis of large LLS networks by means of upward hierarchical analysis," proc. ECCTD, pp. 358-361, 1981.

[CHA_90] Z. Y. Chang, "Low-noise wide-band amplifiers in bipolar and CMOS technologies," PhD dissertation, Katholieke Universiteit Leuven, 1990.

[CHE_66] E. W. Cheney, "Introduction to approximation theory," McGraw-Hill, 1966.

[CHE_87] Y. Cheng, P. M. Lin, "Symbolic analysis of general switched capacitor networks - new methods and implementation," proc. ISCAS, pp. 55-59, 1987.

[CHE_89] D. J. Chen, J.-C. Lee, B. J. Sheu, "SLAM: a smart analog module layout generator for mixed analog-digital VLSI design," proc. ICCD, pp. 24-27, 1989.

[CHO_85] J. T. Chow, A. N. Willson Jr., "A microcomputer-oriented algorithm for symbolic network analysis," proc. ISCAS, pp. 575-577, 1985.

[CHO_90a] U. Choudhury, A. Sangiovanni-Vincentelli, "Use of performance sensitivities in routing of analog circuits," proc. ISCAS, pp. 348-351, 1990.

[CHO_90b] U. Choudhury, A. Sangiovanni-Vincentelli, "Constraint-based channel routing for analog and mixed analog/digital circuits," proc. ICCAD, pp. 198-201, 1990.

[CHR_86] E. Christen, J. Vlach, K. Singhal, M. Vlach, "A new program for analysis and optimization of passive and active-RC networks," proc. ISCAS, pp. 972-975, 1986.

[CHU_75] L. O. Chua, P. M. Lin, "Computer-aided analysis of electronic circuits: algorithms and computational techniques," especially chapter 14, Prentice-Hall, 1975.

[CHU_79] L. O. Chua, C.-Y. Ng, "Frequency-domain analysis of nonlinear systems: formulation of transfer functions," IEE Journal of Electronic Circuits and Systems, Vol. 3, No. 6, pp. 257-269, November 1979.

[CHU_87] L. O. Chua, C. A. Desoer, E. S. Kuh, "Linear and nonlinear circuits," McGraw-Hill, 1987.

[CLA_84]   L. Claesen, "Computer aided design of integrated systems for digital and analog signal processing," PhD dissertation, Katholieke Universiteit Leuven, 1984.

[COC_73]   B. L. Cochrun, A. Grabel, "A method for the determination of the transfer function of electronic circuits," IEEE Transactions on Circuit Theory, Vol. CT-20, No. 1, pp. 16-20, January 1973.

[COH_90]   J. M. Cohn, D. J. Garrod, R. A. Rutenbar, L. R. Carley, "New algorithms for placement and routing of custom analog cells in ACACIA," proc. CICC, pp. 27.6.1-27.6.5, 1990.

[COL_89]   B. C. Cole, "The changing analog world," Electronics, pp. 73-80, September 1989.

[COT_90]   R. A. Cottrell, "Event-driven behavioural simulation of analogue transfer functions," proc. European Design Automation Conference, pp. 240-243, 1990.

[COX_80]   D. B. Cox, L. T. Lin, "A realtime programmable switched capacitor filter," proc. ISSCC, pp. 94-95, 1980.

[CRU_90]   J. Cruz Moreno, "Hierarchical analytic modeling of fully differential amplifier topologies for analog silicon compilation," Master's thesis, Katholieke Universiteit Leuven, 1990.

[DEG_83]   M. Degrauwe, "Studie en realisatie van microvermogen filters met geschakelde capaciteiten," PhD dissertation (in Dutch), Katholieke Universiteit Leuven, 1989.

[DEG_84]   M. G. Degrauwe, W. M. Sansen, "A synthesis program for operational amplifiers," proc. ISSCC, pp. 18-19, 1984.

[DEG_86]   M. Degrauwe, "User's guide of IDAC," CSEM, 1986.

[DEG_87a] M. Degrauwe, O. Nys, E. Vittoz, E. Dijkstra, J. Rijmenants, S. Bitz, S. Cserveny, J. Sanchez, "An analog expert design system," proc. ISSCC, pp. 212-213, 1987.

[DEG_87b] M. Degrauwe, M. Declercq, R. Zinszner, M. Kayal, "CAD tools for analog ASIC's," proc. Journées d'Electronique EPFL, Lausanne, 1987.

[DEG_87c] M. Degrauwe, "IDAC: an interactive design tool for analog CMOS circuits," IEEE Journal of Solid-State Circuits, Vol. SC-22, No. 6, pp. 1106-1116, December 1987.

[DEG_88a] M. Degrauwe, "Analog design automation," Summer Course EPFL on CMOS VLSI Design Analog & Digital, Lausanne, 1988.

[DEG_88b] M. Degrauwe, "Present status and future trends in analog design automation," proc. ESSCIRC, pp. 313-318, 1988.

[DEG_89]   M. Degrauwe, B. Goffart, C. Meixenberger, M. Pierre, J. Litsios, J. Rijmenants, O. Nys, E. Dijkstra, B. Joss, M. Meyvaert, T. Schwarz, M. Pardoen, "Towards an analog system design environment," IEEE Journal of Solid-State Circuits, Vol. SC-24, No. 3, pp. 659-671, June 1989.

[DEG_90]   M. Degrauwe, B. Goffart, B. Joss, J. Rijmenants, C. Meixenberger, T. Schwarz, J. Litsios, S. Seda, M. Pierre, J. Jongsma, C. Scarnera, T. Hornstein, P. Deck, "The ADAM analog design automation system," proc. ISCAS, pp. 820-822, 1990.

[DEK_84]   J. De Kleer, "How circuits work," Artificial Intelligence (Elsevier Science Publishers), No. 24, pp. 205-280, 1984.

[DEL_89]   S. Deloof, L. Nachtergaele, "CASNA: een programma voor het opstellen van symbolische transfertfunkties," Master's thesis (in Dutch), Katholieke Industriële Hogeschool West-Vlaanderen (Oostende), 1989.

[DEM_86]   H. De Man, J. Rabaey, P. Six, L. Claessen, "CATHEDRAL II: a silicon compiler for digital signal processing multiprocessor VLSI systems," IEEE Design & Test of Computers, Vol. 3, No. 6, pp. 13-26, December 1986.

[DEM_90]   H. De Man, "Silicon compilation for real time signal processing systems," tutorial High-level Synthesis part II, European Design Automation Conference, 1990.

[DIR_69]   S. W. Director, R. A. Rohrer, "The generalized adjoint network and network sensitivities," IEEE Transactions on Circuit Theory, Vol. CT-16. No. 3, pp. 318-323, August 1969.

[DIR_90]   S. W. Director, W. Maly, A. Strojwas, "Statistically based IC process simulation and applications in VLSI design and manufacturing," Kluwer Academic Publishers, 1990.

[DON_90]   S. Donnay, "Constructie van een topologieselector voor gebruik in analoge silicon compilatie," Master's thesis (in Dutch), Katholieke Universiteit Leuven, 1990.

[DOW_69]   T. Downs, "Symbolic evaluation of transmittances from the nodal-admittance matrix," Electronics Letters, Vol. 5, No. 16, pp. 379-380, August 1969.

[DOW_70a] T. Downs, "Inversion of the nodal-admittance matrix in symbolic form," Electronics Letters, Vol. 6, No. 3, pp. 74-76, February 1970.

[DOW_70b] T. Downs, "Inversion of nodal-admittance matrix for active networks in symbolic form," Electronics Letters, Vol. 6, No. 22, pp. 690-691, October 1970.

[ELT_86]   F. M. El-Turky, "BLADES: an expert system for analog circuit design," proc. ISCAS, pp. 552-555, 1986.

[ELT_87]   F. M. El-Turky, "A fully automated expert system design environment for operational amplifiers," proc. ECCTD, 1987.

[ELT_89]   F. M. El-Turky, E. E. Perry, "BLADES: an artificial intelligence approach to analog circuit design," IEEE Transactions on Computer-Aided Design, Vol. 8, No. 6, pp. 680-692, June 1989.

[EYC_89]   W. Eyckmans, L. Callewaert, M. Steyaert, W. Sansen, "The Internal Human Conditioning System, a multi purpose programmable biomedical system," Journal of Medical Engineering and Technology, Vol. 13, No. 1/2, pp. 93-95, January/April 1989.

[FAN_80]   S. C. Fang, Y. P. Tsividis, "Modified analysis with improved numerical methods for switched-capacitor networks," proc. ISCAS, pp. 977-980, 1980.

[FAN_83a] S. C. Fang, Y. P. Tsividis, O. Wing, "SWITCAP: a switched-capacitor network analysis program - Part I: basic features," IEEE Circuits and Systems Magazine, Vol. 5, No. 3, pp. 4-10, September 1983.

[FAN_83b] S. C. Fang, Y. P. Tsividis, O. Wing, "SWITCAP: a switched-capacitor network analysis program - Part II: advanced applications," IEEE Circuits and Systems Magazine, Vol. 5, No. 4, pp. 41-46, December 1983.

[FID_73] J. K. Fidler, J. I. Sewell, "Symbolic analysis for computer-aided circuit design - the interpolative approach," IEEE Transactions on Circuit Theory, Vol. CT-20, No. 6, pp. 738-741, November 1973.

[FLE_79] P. E. Fleischer, K. R. Laker, "A family of active switched capacitor biquad building blocks," The Bell System Technical Journal, Vol. 58, No. 10, pp. 2235-2269, December 1979.

[FRI_74] C. Fridas, J. I. Sewell, "Symbolic analysis of networks by partitioned polynomial interpolation," IEEE Transactions on Circuits and Systems, Vol. CAS-21, No. 3, pp. 345-347, May 1974.

[FUN_88] A. H. Fung, D. J. Chen, Y.-N. Lin, B. J. Sheu, "Knowledge-based analog circuit synthesis with flexible architecture," proc. ICCD, pp. 48-51, 1988.

[GAJ_83] D. Gajski, R. Kuhn, "New VLSI tools," Computer, Vol. 16, No. 12, pp. 11-14, December 1983.

[GAJ_88] D. Gajski (editor), "Silicon compilation," Addison Wesley, 1988.

[GAR_79] M. R. Garey, D. S. Johnson, "Computers and intractability - a guide to the theory of NP-completeness," W. H. Freeman, San Francisco, 1979.

[GAR_88] D. J. Garrod, R. A. Rutenbar, L. R. Carley, "Automatic layout of custom analog cells in ANAGRAM," proc. ICCAD, 1988.

[GAT_89] U. Gatti, F. Maloberti, V. Liberali, "Full stacked layout of analogue cells," proc. ISCAS, pp. 1123-1126, 1989.

[GEA_89] M. R. Gearty, J. B. Foley, "OPTIC: a user-oriented optimisation package for IC design," proc. ECCTD, pp. 512-516, 1989.

[GEN_74] W. M. Gentleman, S. C. Johnson, "The evaluation of determinants by expansion by minors and the general problem of substitution," Mathematics of Computation, Vol 28, No. 126, pp. 543-548, April 1974.

[GIE_86] G. Gielen, "Identification of the subtasks for automatic analog design," internal report Katholieke Universiteit Leuven, 1986.

[GIE_89a] G. Gielen, H. Walscharts, W. Sansen, "Analog circuit design optimization based on symbolic simulation and simulated annealing," proc. ESSCIRC, pp. 252-255, 1989.

[GIE_89b] G. Gielen, W. Sansen, "Symbolic calculation of the linear characteristics of analog integrated circuits," proc. BIRA workshop Technical Applications of Automatic Formula Manipulation, Leuven, November 1989.

[GIE_89c] G. Gielen, H. Walscharts, W. Sansen, "ISAAC: a symbolic simulator for analog integrated circuits," IEEE Journal of Solid-State Circuits, Vol. SC-24, No. 6, pp. 1587-1597, December 1989.

[GIE_90a]   G. Gielen, K. Swings, W. Sansen, "An intelligent design system for analogue integrated circuits," proc. European Design Automation Conference, pp. 169-173, 1990.

[GIE_90b]   G. Gielen, H. Walscharts, W. Sansen, "Analog circuit design optimization based on symbolic simulation and simulated annealing," IEEE Journal of Solid-State Circuits, Vol. SC-25, No. 3, pp. 707-713, June 1990.

[GIE_90c]   G. Gielen, W. Sansen, "Automation of analog design procedures," chapter 4 in "Introduction to analog VLSI design automation" (edited by M. Ismail and J. Franca), Kluwer Academic Publishers, 1990.

[GIE_90d]   G. Gielen, "Design automation for analog integrated circuits," PhD dissertation, Katholieke Universiteit Leuven, 1990.

[GOF_89a]   B. Goffart, J. Litsios, M. Degrauwe, "Design strategies versus numerical optimization for analog circuits," proc. ISCAS, pp. 1119-1122, 1989.

[GOF_89b]   B. Goffart, J. Jongsma, M. Degrauwe, "Worst case design and datasheet generation techniques for analog silicon compilers," proc. ESSCIRC, pp. 125-128, 1989.

[GRA_84]   P. R. Gray, R. G. Meyer, "Analysis and design of analog integrated circuits (second edition)," John Wiley & Sons, 1984.

[GRE_84]   A. B. Grebene, "Bipolar and CMOS analog integrated circuit design," John Wiley & Sons, 1984.

[GRE_86]   R. Gregorian, G. C. Temes, "Analog MOS integrated circuits for signal processing," John Wiley & Sons, 1986.

[GRI_76]   M. L. Griss, "An efficient sparse minor expansion algorithm," Proceedings ACM 76, pp. 429-434, 1976.

[GRU_80]   D. von Grünigen, U, W. Brugger, "Closed-form analysis of nonideal switched-capacitor networks," proc. ECCTD, pp. 165-170, 1980.

[GYU_89]   R. S. Gyurcsik, J.-C. Jeen, "A generalized approach to routing mixed analog and digital signal nets in a channel," IEEE Journal of Solid-State Circuits, Vol. SC-24, No. 2, pp. 436-442, April 1989.

[HAB_87]   E. Habekotté, B. Hoefflinger, H.-W. Klein, M. A. Beunder, "State of the art in the analog CMOS circuit design," Proceedings of the IEEE, Vol. 75, No. 6, pp. 816-828, June 1987.

[HAR_84]   D. I. Hariton, "Multiport matrix signal flow graphs: a heuristic, topological method for the minimization of the gain formula," proc. ISCAS, pp. 1407-1410, 1984.

[HAR_87]   R. Harjani, R. A. Rutenbar, L. R. Carley, "A prototype framework for knowledge-based analog circuit synthesis," proc. Design Automation Conference, pp. 42-49, 1987.

[HAR_88a]   R. Harjani, R. A. Rutenbar, L. R. Carley, "Analog circuit synthesis and exploration in OASYS," proc. ICCD, 1988.

[HAR_88b]   R. Harjani, R. A. Rutenbar, L. R. Carley, "Analog circuit synthesis for performance in OASYS," proc. ICCAD, 1988.

[HAR_89a]   R. Harjani, "OASYS: a framework for analog circuit synthesis," PhD dissertation, Carnegie Mellon University, 1989.

[HAR_89b]   R. Harjani, R. A. Rutenbar, L. R. Carley, "OASYS: a framework for analog circuit synthesis," IEEE Transactions on Computer-Aided Design, Vol. 8, No. 12, pp. 1247-1266, December 1989.

[HAS_89a]   M. M. Hassoun, P. M. Lin, "A new network approach to symbolic simulation of large-scale networks," proc. ISCAS, pp. 806-809, 1989.

[HAS_89b]   M. Hashizume, H, Y. Kawai, K. Nii, T. Tamesada, "Design automation system for analog circuits based on fuzzy logic," proc. CICC, pp. 4.6.1-4, 1989.

[HEI_88]   P. Heikkilä, M. Valtonen, K. Mannersalo, "CMOS op amp dimensioning using multiphase optimization," proc. ISCAS, pp. 167-170, 1988.

[HEI_89]   P. Heikkilä, M. Valtonen, H. Pohjonen, "Automated dimensioning of MOS transistors without using topology-specific explicit formulas," proc. ISCAS, pp. 1131-1134, 1989.

[HEN_89]   R. K. Henderson, L. Ping, J. I. Sewell, "A design program for digital and analogue filters: PANDDA," proc. ECCTD, pp. 289-293, 1989.

[HO_75]   C.-W. Ho, A. E. Ruehli, P. A. Brennan, "The modified nodal approach to network analysis," IEEE Transactions on Circuits and Systems, Vol. CAS-22, No. 6, pp. 504-509, June 1975.

[HSU_89]   W.-J. Hsu, B. J. Sheu, "Digital and analog integrated-circuit design with built-in reliability," proc. ICCD, pp. 496-499, 1989.

[HUE_89]   L. P. Huelsman, "Personal computer symbolic analysis programs for undergraduate engineering courses," proc. ISCAS, pp. 798-801, 1989.

[IPS_89]   J. Ipsen, L. Moore, "CELLO: an integrated analog macrocell CAD system," proc. ISCAS, pp. 1135-1140, 1989.

[ISM_90]   M. Ismail, J. Franca, "Introduction to analog VLSI design automation" Kluwer Academic Publishers, 1990.

[JOH_84]   D. G. Johnson, J. I. Sewell, "Some improvement schemes for the interpolative symbolic analysis of switched capacitor networks," proc. ISCAS, pp. 1324-1327, 1984.

[JUS_90]   G. Jusuf, P. R. Gray, A. L. Sangiovanni-Vincentelli, "CADICS - Cyclic analog-to-digital converter synthesis," proc. ICCAD, pp. 286-289, 1990.

[KAM_83]   M. Kameda, A. Niimura, H. Nakayama, I. Iwama, "A new bipolar-CMOS gate array for analog-digital applications," proc. CICC, pp. 189-193, 1983.

[KAW_88]   M. Kawakita, T. Watanabe, "Analog layout compaction with a clean-up function," Transactions of the IEICE, Vol. E 71, No. 12, pp. 1243-1252, December 1988.

[KAY_88a]   M. Kayal, S. Piguet, M. Declercq, B. Hochet, "SALIM: a layout generation tool for analog ICs," proc. CICC, pp. 7.5.1-4, 1988.

[KAY_88b]   M. Kayal, S. Piguet, M. Declercq, B. Hochet, "An interactive layout generation tool for CMOS analog ICs," proc. ISCAS, pp. 2431-2434, 1988.

[KAY_89]   M. Kayal, "Génération du layout de circuits intégrés analogiques assistée par ordinateur," PhD dissertation (in French), Ecole Polytechnique Fédérale de Lausanne, 1989.

[KEL_85]   G. Kelson, "Design automation techniques for analog VLSI," VLSI Design, pp. 78-82, January 1985.

[KEL_88]   T. M. Kelessoglou, D. O. Pederson, "A knowledge-based SPICE environment for improved convergence and user friendliness," proc. CICC, pp. 3.1.1-4, 1988.

[KEL_89]   T. M. Kelessoglou, D. O. Pederson, "NECTAR: a knowledge-based environment to enhance SPICE," IEEE Journal of Solid-State Circuits, Vol. SC-24, No. 2, pp. 452-457, April 1989.

[KEN_88]   J. G. Kenney, L. R. Carley, "CLANS: a high-level synthesis tool for high resolution data converters," proc. ICCAD, 1988.

[KIM_89]   C. K. Kim, E. Berkcan, B. Currin, M. d'Abreu, "A new floorplanning algorithm for analog circuits," proc. CICC, pp. 3.2.1-4, 1989.

[KIN_85]   P. King, "Subcircuits on linear arrays - a new array topology," proc. CICC, pp. 475-478, 1985.

[KIR_83]   S. Kirkpatrick, C. D. Gelatt Jr., M. P. Vecchi, "Optimization by simulated annealing," Science, Vol. 220, No. 4598, pp. 671-680, 13 May 1983.

[KOH_87]   H. Y. Koh, C. H. Séquin, P. R. Gray, "Automatic synthesis of operational amplifiers based on analytic circuit models," proc. ICCAD, pp. 502-505, 1987.

[KOH_88]   H. Y. Koh, C. H. Séquin, P. R. Gray, "Automatic layout generation for CMOS operational amplifiers," proc. ICCAD, pp. 548-551, 1988.

[KOH_89]   H. Y. Koh, "Design synthesis of monolithic CMOS operational amplifiers," PhD dissertation, University of California at Berkeley, 1989.

[KOH_90]   H. Y. Koh, C. H. Séquin, P. R. Gray, "OPASYN: a compiler for CMOS operational amplifiers," IEEE Transactions on Computer-Aided Design, Vol. 9, No. 2, pp. 113-125, February 1990.

[KON_80]   A. Konczykowska, J. Starzyk, "Computer analysis of large signal flowgraphs by hierarchical decomposition method," proc. ECCTD, pp. 408-413, 1980.

[KON_81]   A. Konczykowska, J. Starzyk, "Computer justification of the upward topological analysis of flow-graphs," proc. ECCTD, pp. 464-467, 1981.

[KON_88]   A. Konczykowska, M. Bon, "Automated design software for switched-capacitor IC's with symbolic simulator SCYMBAL," proc. Design Automation Conference, pp. 363-368, 1988.

[KON_89]   A. Konczykowska, M. Bon, "Symbolic simulation for efficient repetitive analysis and artificial intelligence techniques in C.A.D.," proc. ISCAS, pp. 802-805, May 1989.

[KOS_89]   A. P. Kostelijk, "VERA: a rule-based verification assistant for VLSI circuit design," proc. VLSI'89, pp. 89-98, 1989.

[KUH_87]   J. Kuhn, "Analog module generators for silicon compilation," VLSI Systems Design, pp. 74-80, May 1987.

[KUN_86a] K. S. Kundert, "Sparse matrix techniques," chapter 6 of "Circuit analysis, simulation and design" (editor Ruehli), Elsevier Science Publishers (North Holland), pp. 281-324, 1986.

[KUN_86b] K. S. Kundert, A. Sangiovanni-Vincentelli, "Simulation of nonlinear circuits in the frequency domain," IEEE Transactions on Computer-Aided Design, Vol. CAD-5, No. 4, pp. 521-535, October 1986.

[KUN_88] K. S. Kundert, E. Copeland, A. Sangiovanni-Vincentelli, "SPECTRE user's guide. A frequency-domain simulator for nonlinear circuits. Version 1a1," Department of Electrical Engineering and Computer Sciences, University of California, Berkeley, 1988.

[KUR_90] C. M. Kurker, J. J. Paulos, B. S. Cohen, E. S. Cooley, "Development of an Analog Hardware Description Language," proc. CICC, pp. 5.4.1-5.4.6, 1990.

[LAI_88] J. C. Lai, J. S. Kueng, H. J. Chen, F. J. Fernandez, "ADOPT - a CAD system for analog circuit design," proc. CICC, pp. 3.2.1-4, 1988.

[LAK_91] K. Laker, W. Sansen, "Design of analog integrated circuits and systems", McGraw-Hill, 1991.

[LEE_74] A. Y. Lee, "Signal flow graphs - computer-aided system analysis and sensitivity calculations," IEEE Transactions on Circuits and Systems, Vol. CAS-21, No. 2, pp. 209-216, March 1974.

[LEE_80] B. G. Lee, "The product matrices and new gain formulas," IEEE Transactions on Circuits and Systems, Vol. CAS-27, No. 4, pp. 284-292, April 1980.

[LEE_90] D. M. W. Leenaerts, "Application of interval analysis for circuit design," IEEE Transactions on Circuits and Systems, Vol. CAS-37, No. 6, pp. 803-807, June 1990.

[LI_85] F. Li, "K-tree topological method for producing symbolic sensitivity functions," proc. ISCAS, pp. 805-806, 1985.

[LIB_88] A. Liberatore, S. Manetti, "SAPEC - a personal computer program for the symbolic analysis of electric circuits," proc. ISCAS, pp. 897-900, 1988.

[LIB_89] A. Liberatore, S. Manetti, "Network sensitivity analysis via symbolic formulation," proc. ISCAS, pp. 705-708, 1989.

[LIN_73a] P. M. Lin, "Computer generation of symbolic network functions - An overview," Computer-aided design (proc. of the IFIP working conference on principles of CAD (October 1982), edited by Vlietstra and Wielinga), North-Holland, pp. 261-282, 1973.

[LIN_73b] P. M. Lin, "A survey of applications of symbolic network functions," IEEE Transactions on Circuit Theory, Vol. CT-20, No. 6, pp. 732-737, November 1973.

[MAI_89] K. Maio, S. Hayashi, M. Furihata, S. Ogura, T. Watanabe, M. Ishikawa, "A highly efficient design system for mixed analog/digital LSIs," proc. ESSCIRC, pp. 121-124, 1989.

[MAL_90] E. Malavasi, U. Choudhury, A. Sangiovanni-Vincentelli, "A routing methodology for analog integrated circuits," proc. ICCAD, pp. 202-205, 1990.

[MAS_53] S. J. Mason, "Feedback theory - some properties of signal flow graphs," Proceedings of the I.R.E., pp. 1144-1156, September 1953.

[MAT_84] T. Matsumoto, M. Kitai, Y. Ohta, "Topological generation of symbolic network functions for general linear networks by signal flowgraph approach," proc. ISCAS, pp. 1403-1406, 1984.

[MEA_89] C. Mead, "Analog VLSI and neural systems," Addison-Wesley, 1989.

[MEI_88a] C. Meixenberger, B. Goffart, M. Pierre, M. Degrauwe, "Sizing algorithms for linear analog circuits," proc. ESSCIRC, pp. 190-193, 1988.

[MEI_88b] C. Meixenberger, M. Degrauwe, "A design program for comparators," proc. ESSCIRC, pp. 198-201, 1988.

[MIE_78] R. R. Mielke, "A new signal flowgraph formulation of symbolic network functions," IEEE Transactions on Circuits and Systems, Vol. CAS-25, No. 6, pp. 334-340, June 1978.

[MIL_90] L. Milor, A. Sangiovanni-Vincentelli, "Optimal test set design for analog circuits," proc. ICCAD, pp. 294-297, 1990.

[MOS_84] G. S. Moschytz, U. W. Brugger, "Signal-flow graph analysis of SC networks," IEE Proceedings part G, Vol. 131, No. 2, pp. 72-85, April 1984.

[MOS_86] G. S. Moschytz, J. J. Mulawka, "New methods of direct closed-form analysis of switched-capacitor networks," proc. ISCAS, pp. 369-372, 1986.

[MOS_89] G. S. Moschytz, A. Dobrowski, "By-inspection signal-flow graph analysis of general switched-capacitor networks," proc. ECCTD, pp. 256-260, 1989.

[MOU_83] Z. Mou-yan, "A network extension algorithm for generating symbolic network functions," proc. ISCAS, pp. 641-644, 1983.

[NAG_71] L. Nagel, R. Rohrer, "Computer analysis of nonlinear circuits excluding radiation (CANCER)," IEEE Journal of Solid-State Circuits, Vol. SC-6, No. 4, pp. 166-182, August 1971.

[NEG_90] M. Negahban, D. Gajski, "Silicon compilation of switched-capacitor networks," proc. European Design Automation Conference, pp. 164-168, 1990.

[NYE_81] W. Nye, E. Polak, A. Sangiovanni-Vincentelli, A. Tits, "DELIGHT: an optimization-based computer-aided design system," proc. ISCAS, pp. 851-855, 1981.

[NYE_88] W. Nye, D. C. Riley, A. Sangiovanni-Vincentelli, A. L. Tits, "DELIGHT.SPICE: an optimization-based system for the design of integrated circuits," IEEE Transactions on Computer-Aided Design, Vol. 7, No. 4, pp. 501-519, April 1988.

[OCH_89] E. S. Ochotta, "The OASYS Virtual Machine: formalizing the OASYS analysis synthesis framework," Master's thesis, Carnegie Mellon University, 1989.

[OHT_86] T. Ohtsuki (editor), "Layout design and verification," Advances in CAD for VLSI series, Volume 4, North-Holland, 1986.

[ONO_89]   H. Onodera, H. Kanbara, K. Tamaru, "Operational amplifier compilation with performance optimization," proc. CICC, pp. 17.4.1-6, 1989.

[ONO_90]   H. Onodera, H. Kanbara, K. Tamaru, "Operational-amplifier compilation with performance optimization," IEEE Journal of Solid-State Circuits, Vol. SC-25, No. 2, pp. 466-473, April 1990.

[OPT_90]   F. Op 't Eynde, "High-performance analog interfaces for digital signal processors," PhD dissertation, Katholieke Universiteit Leuven, 1990.

[OTT_83]   R. Otten, "Efficient floorplan optimization," proc. ICCD, pp. 499-502, 1983.

[OZG_73]   U. Ozguner, "Signal-flowgraph analysis using only loops," Electronics Letters, Vol. 9, No. 16, pp. 359-360, August 1973.

[PET_90]   W. Van Petegem, D. De Wachter, W. Sansen, "Electrothermal simulation of integrated circuits," proc. IEEE Semiconductor Thermal and Temperature Measurement Symposium, pp. 70-73, 1990.

[PIC_83]   T. W. Pickerell, "New analog capabilities on semi-custom CMOS," proc. CICC, pp. 174-179, 1983.

[PIE_84]   M. Pierzchala, "Corrected gain formulas," IEEE Transactions on Circuits and Systems, Vol. CAS-31, No. 6, pp. 581-582, June 1984.

[PIG_89]   S. Piguet, F. Rahali, M. Declercq, M. Kayal, "An analog-oriented routing tool for CMOS analog integrated circuits," proc. ESSCIRC, pp. 80-83, 1989.

[PIG_90]   S. Piguet, F. Rahali, M. Kayal, E. Zysman, M. Declercq, "A new routing method for full custom analog IC's," proc. CICC, pp. 27.7.1-27.7.4, 1990.

[PLE_86]   T. Pletersek, J. Trontelj, L. Trontelj, I. Jones, G. Shenton, "High-performance designs with CMOS analog standard cells," IEEE Journal of Solid-State Circuits, Vol. SC-21, No. 2, pp. 215-222, April 1986.

[PLI_89]   J. O. Pliam, "Ring graphs and gain formulas, an algebraic approach to topology," proc. ISCAS, pp. 327-330, 1989.

[POR_89]   F. Pörnbacher, "A new method supporting the nominal design of analog integrated circuits with regard to constraints," proc. ECCTD, pp. 614-618, 1989.

[POW_81]   M. J. D. Powell, "Approximation theory and methods," Cambridge University Press, 1981.

[PUN_85]   C. K. Pun, J. I. Sewell, "Symbolic analysis of ideal and non-ideal switched capacitor networks," proc. ISCAS, pp. 1165-1168, 1985.

[RAA_88]   R. R. M. van Raaij, "An analog synthesis tool," Philips Natuurkundig Laboratorium Technical Note No. 268/88.

[RAB_83]   J. Rabaey, "A unified computer aided design technique for switched capacitor systems in the time and the frequency domain," PhD dissertation, Katholieke Universiteit Leuven, 1983.

[RES_84]   A. L. Ressler, "A circuit grammer for operational amplifier design," PhD dissertation, Massachusetts Institute of Technology, 1984.

[RIE_72]   D. E. Riegle, P.M. Lin, "Matrix signal flow graphs and an optimum topological method for evaluating their gains," IEEE Transactions on Circuit Theory, Vol. CT-19, No. 5, pp. 427-435, September 1972.

[RIJ_88]   J. Rijmenants, T. Schwarz, J. Litsios, R. Zinszner, "ILAC: an automated layout tool for analog CMOS circuits," proc. CICC, pp. 7.6.1-4, 1988.

[RIJ_89]   J. Rijmenants, J. B. Litsios, T. R. Schwarz, M. G. R. Degrauwe, "ILAC: an automated layout tool for analog CMOS circuits," IEEE Journal of Solid-State Circuits, Vol. SC-24, No. 2, pp. 417-425, April 1989.

[ROH_90]   R. A. Rohrer, personal communication, 1990.

[ROY_80]   G. L. Roylance, "A simple model of circuit design," Master's thesis, Massachusetts Institute of Technology, 1980.

[RUE_86]   A. E. Ruehli (editor), "Circuit analysis, simulation and design," Advances in CAD for VLSI series, Volume 3, Part 2, North-Holland, 1987.

[SAB_90]   "SABER: simulation for circuits through systems," Analogy Inc., 1990.

[SAN_77]   A. Sangiovanni-Vincentelli, L.-K. Chen, L. O. Chua, "An efficient heuristic cluster algorithm for tearing large-scale networks," IEEE Transactions on Circuits and Systems, Vol. CAS-24, No. 12, pp. 709-717, December 1977.

[SAN_80]   P. Sannuti, N. N. Puri, "Symbolic network analysis - an algebraic formulation," IEEE Transactions on Circuits and Systems, Vol. CAS-27, No. 8, pp. 679-687, August 1980.

[SAN_87]   W. Sansen, "Analog integrated circuits with high dynamic range," proc. ESSCIRC, pp. 93-100, 1987.

[SAN_88a]   W. Sansen, "Automatization of analogue design procedures," presymposium on Expert System Tools for Analog Signal Processing, ISCAS, 1988.

[SAN_88b]   W. Sansen, G. Gielen, F. Op 't Eynde, "Operational amplifier design," COMETT report 643 D, November 1988.

[SAN_89]   W. Sansen, G. Gielen, H. Walscharts, "A symbolic simulator for analog circuits", proc. ISSCC, pp. 204-205, February 1989.

[SAN_90]   B. Santo, K. T. Chen, "Technology '90: solid state," IEEE Spectrum, pp. 41-43, January 1990.

[SAS_82]   T. Sasaki, H. Murao, "Efficient gaussian elimination method for symbolic determinants and linear systems," ACM Transactions on Mathematical Software, Vol. 8, No. 3, pp. 277-289, September 1982.

[SAS_89]   D. H. Sassene, "Analog CMOS design tools and chips," Fourth NORCHIP/NORSILC seminar, 1989.

[SCH_80]   M. Schetzen, "The Volterra and Wiener theories of nonlinear systems," John Wiley & Sons, 1980.

[SCH_88]   K. Van Schuylenbergh, R. Verbeeck, "Layouthulpmiddelen voor analoge silicon compiler," Master's thesis (in Dutch), Katholieke Universiteit Leuven, 1988.

[SEC_85]   C. Sechen, A. Sangiovanni-Vincentelli, "The TimberWolf placement and routing package," IEEE Journal of Solid-State Circuits, Vol. SC-20, No. 2, pp. 510-522, April 1985.

[SED_88]   S. J. Seda, M. G. R. Degrauwe, W. Fichtner, "A symbolic analysis tool for analog circuit design automation," proc. ICCAD, pp. 488-491, 1988.

[SED_89]   S. J. Seda, personal communication, visit in University of Zurich, September 1989.

[SHE_88]   B. J. Sheu, A. H. Fung, Y.-N. Lai, "A knowledge-based approach to analog IC design," IEEE Transactions on Circuits and Systems, Vol. CAS-35, No. 2, pp. 256-258, February 1988.

[SHI_74]   S.-D. Shieu, S.-P. Chan, "Topological formulation of symbolic network functions and sensitivity analysis of active networks," IEEE Transactions on Circuits and Systems, Vol. CAS-21, No. 1, pp. 39-45, January 1974.

[SHI_86]   H. Shin, A. Sangiovanni-Vincentelli, "MIGHTY: a rip-up and reroute detailed router," proc. ICCD, pp. 10-13, 1986.

[SHY_88]   J.-M. Shyu, A. Sangiovanni-Vincentelli, "ECSTASY: a new environment for IC design optimization," proc. ICCAD, pp. 484-487, 1988.

[SIG_87]   R. P. Sigg, A. Kaelin, A. Muralt, W. C. Black Jr., G. S. Moschytz, "An SC filter compiler: fully automated filter synthesizer and mask generator for a CMOS gate-array-type filter chip," proc. ICCAD, pp. 510-513, 1987.

[SIN_74]   K. Singhal, J. Vlach, "Generation of immitance functions in symbolic form for lumped distributed active networks," IEEE Transactions on Circuits and Systems, Vol. CAS-21, No. 1, pp. 57-67, January 1974.

[SIN_77]   K. Singhal, J. Vlach, "Symbolic analysis of analog and digital circuits," IEEE Transactions on Circuits and Systems, Vol. CAS-24, No. 11, pp. 598-609, November 1977.

[SIN_86]   K. Singhal, J. Vlach, "Formulation of circuit equations," chapter 2 of "Circuit analysis, simulation and design" (editor Ruehli), Elsevier Science Publishers (North Holland), pp. 45-70, 1986.

[SIU_85]   W.-C. Siu, C.-F. Chen, "A new technique for symbolic function generation using number theoretic transform," proc. ISCAS, pp. 1525-1526, 1985.

[SMI_76]   J. Smit, "The efficient calculation of symbolic determinants," proc. SYMSAC, pp. 105-113, 1976.

[SMI_78a] J. Smit, "An efficient factoring symbolic determinant expansion algorithm," Internal Report of the Twente University of Technology, THT 1231-AM-0478, 1978.

[SMI_78b] J. Smit, "A recursive tearing technique for systems in which small as well as large element values are significant," Internal Report of the Twente University of Technology, THT 1231-AM-0378, 1978.

[SMI_79]   J. Smit, "New recursive minor expansion algorithms, a presentation in a comparative context," proc. EUROSAM, Lecture Notes in Computer Science No. 72 (Springer-Verlag), pp. 74-87, 1979.

[SMI_81a]   J. Smit, "Sparse Kirchhoff equations, an effective support tool for the numeric and symbolic solution of large sparse systems of network equations," proc. ECCTD, pp. 352-357, 1981.

[SMI_81b]   J. Smit, "A cancellation-free algorithm, with factoring capabilities, for the efficient solution of large sparse sets of equations," proc. SYMSAC, pp. 146-154, 1981.

[SMI_82]   J. Smit, J. A. van Hulzen, "Symbolic-numeric methods in microwave technology," proc. EUROCAM, Springer Verlag, pp. 281-288, April 1982.

[SMI_89a]   L. D. Smith, H. R. Farmer, M. Kunesh, M. A. Massetti, D. Willmott, R. Hedman, R. Richetta, T. J. Schmerbeck, "A CMOS-based analog standard cell product family," IEEE Journal of Solid-State Circuits, Vol. SC-24, No. 2, pp. 370-379, April 1989.

[SMI_89b]   M. J. S. Smith, C. Anagnostopoulos, C. Portmann, R. Rao, P. Valdenaire, H. Ching, "Construction of analog library cells for analog/digital ASICs using novel design and modular assembly techniques," proc. CICC, pp. 25.6.1-5, 1989.

[SMI_89c]   M. J. S. Smith, C. Portmann, C. Anagnostopoulos, P. S. Tschang, R. Rao, P. Valdenaire, H. Ching, "Cell libraries and assembly tools for analog/digital CMOS and BiCMOS application-specific integrated circuit design," IEEE Journal of Solid-State Circuits, Vol. SC-24, No. 5, pp. 1419-1432, October 1989.

[STA_80a]   J. Starzyk, "Signal-flow-graph analysis by decomposition method," IEE Proceedings part G, Vol. 127, No. 2, pp. 81-86, April 1980.

[STA_80b]   J. Starzyk, E. Sliwa, "Hierarchic decomposition method for the topological analysis of electronic networks," Circuit Theory and Applications (John Wiley), Vol. 8, pp. 407-417, 1980.

[STA_84]   J. Starzyk, E. Sliwa, "Upward topological analysis of large circuits using directed graph representation," IEEE Transactions on Circuits and Systems, Vol. CAS-31, No. 4, pp. 410-414, April 1984.

[STA_86]   J. A. Starzyk, A. Konczykowska, "Flowgraph analysis of large electronic networks," IEEE Transactions on Circuits and Systems, Vol. CAS-33, No. 3, pp. 302-315, March 1986.

[STA_89]   J. Starzyk, E. Sliwa, "Tolerances in symbolic network analysis," proc. ISCAS, pp. 810-813, May 1989.

[STE_84]   G. L. Steele Jr., "COMMON LISP : the language," Digital Press, 1984.

[STE_87]   M. Steyaert, "Monolithic low-power data-acquisition system for biomedical purpose," PhD dissertation, Katholieke Universiteit Leuven, 1987.

[STI_90]   M. T. van Stiphout, "PLATO - a piecewise linear analysis tool for mixed-level circuit simulation," PhD dissertation, Technische Universiteit Eindhoven, 1990.

[STO_84]   D. C. Stone, J. E. Schroeder, R. H. Kaplan, A. R. Smith, "Analog CMOS building blocks for custom and semicustom applications," IEEE Journal of Solid-State Circuits, Vol. SC-19, No. 1, pp. 55-61, February 1984.

[STO_86]   H. Stoffels, E. H. Nordholt, "Automated noise optimization in amplifiers," proc. ISCAS, pp. 1139-1141, 1986.

[STO_88a] J. Stoffels, "Automated frequency compensation in high-performance amplifiers," proc. Midwest Symposium on Circuits and Systems, pp. 223-226, 1988.

[STO_88b] J. Stoffels, "Automation in high-performance negative feedback amplifier design," PhD dissertation, Technische Universiteit Delft, 1988.

[SWA_83]   "SWAP 2.2 reference manual," Silvar Lisco, Heverlee, Belgium, Doc. M-037-2, 1983.

[SWI_89]   K. Swings, "An intelligent analog design methodology for nonroutine design of analog circuits," internal report Katholieke Universiteit Leuven, 1989.

[SWI_90]   K. Swings, G. Gielen, W. Sansen, "An intelligent analog IC design system based on manipulation of design equations," proc. CICC, pp. 8.6.1-8.6.4, 1990.

[TAL_65]   A. Talbot, "Topological analysis of general linear networks," IEEE Transactions on Circuit Theory, pp. 170-180, June 1965.

[TAL_66]   A. Talbot, "Topological analysis for active networks," IEEE Transactions on Circuit Theory, pp. 111-112, March 1966.

[TAN_82]   M. Tanaka, S. Mori, "Topological formulations for the coefficient matrices of state equations for switched-capacitor networks," IEEE Transactions on Circuits and Systems, Vol. CAS-29, No. 2, pp. 106-115, February 1982.

[THE_82]   Y. Therasse, P. Guebels, P. Jespers, "An automatic CAD tool for switched capacitor filters design: a method based on the generalized Orchard argument," proc. ESSCIRC, 1982.

[THE_87]   Y. Therasse, L. Reynders, R. Lannoo, B. Dupont, "A switched-capacitor filter compiler," VLSI Systems Design, pp. 85-88, September 1987.

[TOU_89]   C. Toumazou, C. A. Makris, C. M. Berrah, P. Y. K. Cheung, "A methodology for automated generation of analogue integrated circuits," proc. ECCTD, pp. 624-628, 1989.

[TRO_87]   L. Trontelj, J. Trontelj, T. Slivnik Jr., R. Sosic, D. Strle, "Analog silicon compiler for switched capacitor circuits," proc. ICCAD, pp. 506-509, 1987.

[TRO_89a] J. Trontelj, L. Trontelj, T. Slivnik, T. Pletersek, "Automatic circuit and layout design for mixed analog/digital ASICs," proc. CICC, pp. 17.1.1-4, 1989.

[TRO_89b] J. Trontelj, L. Trontelj, G. Shenton, "Analog digital ASIC design," McGraw-Hill, 1989.

[TSA_77]   M. K. Tsai, B. A. Shenoi, "Generation of symbolic network functions using computer software techniques," IEEE Transactions on Circuits and Systems, Vol. CAS-24, No. 6, pp. 344-346, June 1977.

[TSI_85]   Y. Tsividis, P. Antognetti (editors), "Design of MOS VLSI circuits for telecommunications," Prentice-Hall, 1985.

[TSI_87]   Y. P. Tsividis, "Analog MOS integrated circuits - Certain new ideas, trends, and obstacles," IEEE Journal of Solid-State Circuits, Vol. SC-22, No. 3, pp. 317-321, June 1987.

[VAN_81]   J. Vandewalle, H. J. De Man, J. Rabaey, "Time, frequency, and z-domain modified nodal analysis of switched-capacitor networks," IEEE Transactions on Circuits and Systems, Vol. CAS-28, No. 3, pp. 186-195, March 1981.

[VAN_83]   J. Vandewalle, J. Rabaey, W. Vercruysse, H. J. De Man, "Computer-aided distortion analysis of switched-capacitor filters in the frequency domain," IEEE Journal of Solid-State Circuits, Vol. SC-18, No. 3, pp. 324-333, June 1983.

[VER_88]   C. Verdonck, G. Van Hecke, "Dimensiegeneratie voor analoge silicon compiler," Master's thesis (in Dutch), Katholieke Universiteit Leuven, 1988.

[VHD_87]   "VHDL Language Reference Manual", ANSI/IEEE standard 1076-1987, IEEE Computer Society Press, 1987.

[VLA_80a] A. Vladimirescu, S. Liu, "The simulation of MOS integrated circuits using SPICE 2," Memorandum No. UCB/ERL M80/7, Electronics Research Laboratory, College of Engineering, University of California, Berkeley, February 1980.

[VLA_80b] A. Vladimirescu, A. R. Newton, D. O. Pederson, "SPICE version 2G.1 user's guide," Department of Electrical Engineering and Computer Sciences, University of California, Berkeley, October 1980.

[VLA_82]   J. Vlach, K. Singhal, M. Vlach, "Analysis of switched-capacitor networks," proc. ISCAS, pp. 9-12, 1982.

[VOS_88]   F. Vos, C. Trullemans, "The symbolic simulator SYMSIM," proc. COMPEURO, 1988.

[VOS_89]   F. Vos, C. Trullemans, "A switch level symbolic simulator," proc. ESSCIRC, pp. 244-247, 1989.

[WAH_89]   B. W. Wah, M. B. Lowrie, G.-J. Li, "Computers for symbolic processing," Proceedings of the IEEE, Vol. 77, No. 4, pp. 509-540, April 1989.

[WAL_87]   H. Walscharts, L. Kustermans, W. Sansen, "Noise optimization of switched-capacitor biquads," IEEE Journal of Solid-State Circuits, Vol. SC-22, No. 3, pp. 445-447, June 1987.

[WAL_89]   H. Walscharts, G. Gielen, W. Sansen, "Symbolic simulation of analog circuits in s- and z-domain," proc. ISCAS, pp. 814-817, May 1989.

[WAL_91]   H. Walscharts, "Silicon compilation for switched capacitor filters," PhD dissertation, Katholieke Universiteit Leuven, 1990.

[WAM_90]   P. Wambacq, G. Gielen, W. Sansen, "Symbolic simulation of harmonic distortion in analog integrated circuits with weak nonlinearities," proc. ISCAS, pp. 536-539, 1990.

[WAM_91]   P. Wambacq, J. Vanthienen, G. Gielen, W. Sansen, "A design tool for weakly nonlinear analog integrated circuits with multiple inputs (mixers, multipliers)," proc. CICC, 1991.

[WAN_77]   P. Wang, "On the expansion of sparse symbolic determinants," proc. International Conference on System Sciences, Honolulu, 1977.

[WAN_86]   C.-K. Wang, R. Castello, P. R. Gray, "A scalable high-performance switched-capacitor filter," IEEE Journal of Solid-State Circuits, Vol. SC-21, No. 1, pp. 57-64, February 1986.

[WAW_89]   E. Wawryn, "PROLOG-based active filter synthesis," proc. ECCTD, pp. 670-673, 1989.

[WEB_89]   S. Weber, "Analogue gets automated," Electronics, pp. 62-63, November 1989.

[WIN_87]   G. Winner, T. A. Nguyen, C. Siemaker, "Analog macrocell assembler," VLSI Systems Design, pp. 68-71, May 1987.

[WON_86]   D. F. Wong, C. L. Liu, "A new algorithm for floorplan design," proc. Design Automation Conference, pp. 101-107, 1986.

[XIA_85]   Z. Xiao-Feng, S. Zhi-Guang, "Signal flow graph analysis of switched-capacitor networks," pp. 179-189, 1985.

[YAG_86]   H. Yaguthiel, A. Sangiovanni-Vincentelli, P. R. Gray, "A methodology for automated layout of switched-capacitor filters," proc. ICCD, pp. 444-447, 1986.

[YEU_84]   K. S. Yeung, "Symbolic network function generation via discrete Fourier transform," IEEE Transactions on Circuits and Systems, Vol. CAS-31, No. 2, pp. 229-231, February 1984.

[ZHA_83a] Y.-C. Zhao, "A method for reducing multiplications and divisions in the gain expression of signal flow graphs," proc. ISCAS, pp. 630-632, 1983.

[ZHA_83b] Y.-C. Zhao, "An improved method for reducing multiplications in the gain expression of acyclic signal flow graph," IEEE Transactions on Circuits and Systems, Vol. CAS-30, No. 11, pp. 838-841, November 1983.

INDEX

Engin

TK7874.G54 1991
  Gielen, Georges.
   Symbolic analysis for
automated design of analog
integrated circuits.

3-3-92

ENGINEERING